DTS-34-FA22A-LR1
Federal Aviation Administration
Office of Environment and Energy
Washington, D.C. 20591

A REVIEW OF LITERATURE ON PARTICULATE MATTER

EMISSIONS FROM AIRCRAFT

Roger L. Wayson
Gregg G. Fleming
Brian Kim
U.S. Department of Transportation
Research and Special Programs Administration
John A. Volpe National Transportation Systems Center
Air Quality Facility, DTS-34, Kendall Square
Cambridge, MA 02142-1093

Draft Letter Report

April 2002

U.S. Department of Transportation
Federal Aviation Administration

EXECUTIVE SUMMARY

The Air Quality Facility of the John A. Volpe National Transportation Systems Center is providing support to the Emissions Division of the Federal Aviation Administration's Office of Environment and Energy (AEE-300). As part of this support, the Air Quality Facility is conducting a comprehensive review of past research in the area of aviation-related particulate matter. The purpose of this literature review is to allow an informed decision to be made on a possible first-order approximation to predict the mass of particulate emissions from aircraft until such time that measured data are available for most aircraft. Available references included in this report were collected from public and private libraries, technical journals, project reports, electronic literature searches, interviews, and other key sources. It should be noted that the literature review is not all-inclusive and that emphasis was placed on information concerning measured mass data from aircraft. Following the literature findings, a first order approximation methodology is suggested to estimate the mass of particulate matter (PM) emitted from aircraft based on available data.

Important findings of the literature review include:

- Small PM may be a health concern.
- Most PM emitted by modern transport aircraft has an aerodynamic diameter of less than 2.5 micrometers. This is an important concern considering the EPA health-based standards for $PM_{2.5}$ and PM_{10}.
- The EPA PM standards are mass based (mass/volume of air) but the most complete data base for transport-category aircraft is the ICAO database which reports the smoke number. The smoke number does not correlate well with mass emissions due to the nature of the test procedure used. As such, there is a lack of measured mass data to assist in the analysis to determine if an airport is in compliance with the EPA standards.
- PM are irregular in shape and often coagulate. This coagulation process results in different PM characteristics for different age plumes. This leads to a bi-modal distribution. A lognormal distribution is still appropriate for the soot component (non-volatile PM primarily containing carbon).
- PM include both volatile and non-volatile components. Soot is the most prevalent, non-volatile component. Metals are emitted, but in extremely small amounts.
- Effects on PM emission indices include fuel flow, engine design / operating conditions, altitude, and fuel composition.
- Approximation methods used in PM analysis include: simple factor, compound factor, grab samples or nearby deposition measurements, and measurement mass emission indexes.

The first-order approximation method suggested by the authors is based on several key considerations. These include:

- Airport operators need to account for changes in fleet mix, aircraft modes (related to throttle settings), and airport altitudes to be considered. The simple approximation method and the grab-sample / deposition methods do not permit this flexibility and as such, do not meet current requirements.
- The accuracy of each possible method and the availability of data also were heavily weighed when considering this approximation method. It is a foregone conclusion by the authors that measured data would be more accurate than estimation techniques. However, insufficient information exists now or in the near future to characterize the total fleet and use this method directly.
- The only comprehensive database now available is the ICAO listing of smoke numbers, which are not well related to mass emissions.
- The compound factor approach has been often used by the airport community and could provide the short-term first order approximation that is needed. The largest source of error in this method has always been a connection with mass emissions and smoke number. To help reduce this source of error, the compound factor method must use an adaptation of methodologies that have been derived based on measured data to provide a more accurate correlation to the smoke number.

The suggested first-order approximation method is a combination of the methodologies put forward by the University of Missouri Rolla and the German research agency, DLR. This combined method should allow a more accurate factor to be derived for use in the compound factor method. Measured data would allow a relationship between mass emissions and a small number engine types to be developed. These relationships would then be related to fuel flow to permit an emission factor to be developed. The emission factor would be aircraft specific. Data from the ICAO database would then be used to allow approximation of the mass emissions for most aircraft engine types. The derived mass-based factor should be more accurate than those that have been used in the past.

In the short term, the use of other measured data, such as the applicable engines that have been tested by the Armed Forces, could be used as a calibration of the derived method. In the longer term, as more measured data becomes available, the model could be further calibrated and divided into further aircraft categories until the data base is extensive enough to abandon the approximation method altogether.

TABLE OF CONTENTS

LIST OF FIGURES

LIST OF TABLES

6

1.0 INTRODUCTION

The Air Quality Facility of the John A. Volpe National Transportation Systems Center (Volpe Center) is providing support to the Emissions Division of the Federal Aviation Administration's Office of Environment and Energy (AEE-300). As part of this support, the Air Quality Facility is conducting a comprehensive review of past research in the area of aviation-related particulate matter (PM). In support of this effort, the Center will also stay abreast of current research and make recommendations on future research and related activities.

The purpose of this literature review was to allow an informed decision to be made on a possible first order approximation to predict particulate emissions from aircraft until such time that sufficient measured data are available for most aircraft and approximate methods are no longer needed. The literature review presents a summary of information to be used as the basis for the first-order approximation method.

This letter report first discusses available references that have been collected from libraries, technical journals, project reports, personal libraries, electronic literature searches, interviews, and other key sources. It should be noted that the literature review is not all-inclusive and that emphasis was placed on measured mass data from aircraft. Following the literature findings, a first order approximation is suggested that could be used to estimate the mass of PM emitted from most transport-category aircraft based on available data.

2.0 PROPERTIES OF PARTICULATE MATTER

The gaseous criteria pollutants are relatively well defined but the same cannot be said of air-borne PM. Air-borne PM consists of a broad class of chemically and physically diverse substances, and PM may be classified as a solid or liquid. In the literature, various terms are used for these different types of PM. These include:

- soot or carbon black, an agglomeration of carbon particles;
- aerosol, a dispersion of microscopic solid or liquid particles in a gaseous medium;
- fog, a visible aerosol;
- fume, particles formed by condensation, sublimation, or chemical reaction;
- mist, dispersion of small liquid droplets of sufficient size to settle; and,
- smoke, small gasborne particles resulting from combustion.

PM may be directly released from the source (primary) or form in the ambient air from precursors (secondary). As such, air-borne PM is a subset of all atmospheric aerosols. In sufficient concentration, PM has been linked by epidemiologic studies to "......
mortality, hospital admissions, respiratory symptoms and illness measured in community surveys, and changes in pulmonary mechanical function." [EPA, 1996]. Reduced

7

visibility, effects on climate, detrimental effects on plant life, and material damage have also been reported. This has led to increased public concern and the promulgation of regulations to reduce ambient levels of PM.

2.1 Health Effects and Applicable Standards

Concerns on the effects due to PM have been documented with early studies beginning in the 1930s [EPA, 1996]. The extent of any effects depends on the particle size distribution, exposure dose, and the physiologic status of any individual. It has also been demonstrated that smaller particles present a greater relative risk to the general population.

Sections 108 and 109 of the Clean Air Act (CAA) govern the establishment, review and revision of the U.S. National Ambient Air Quality Standards (NAAQS). These standards were originally established for seven criteria pollutants but now only include six categories of pollutants; one being PM. The U.S. Environmental Protection Agency (EPA) relies on the latest scientific information to set these health-based standards. As such, the PM federal standard has changed over time.

Originally, total suspended particulate (TSP) was regulated. On April 30, 1971 [FR, 1971], EPA promulgated the original primary (to protect human health) and secondary (to protect public welfare) standards. The reference method for measurement was the High Volume sampler (Hi-Vol) [CFR, 1986]. The Hi-Vol collected PM up to about 45 micrometers in diameter and the standard for TSP was 260 micrograms-per-cubic-meter ($\mu g/m^3$) for a 24-hour average, not to be exceeded once-per-year, and 75 $\mu g/m^3$, annual geometric mean. The secondary 24-hour standard was set at 150 $\mu g/m^3$. However, this standard penalized areas subject to prevalent natural dust. Also, health effects are much more significant for smaller PM. As such, the health-related standard was published in the Federal Register in 1987 [FR, 1987] and included only PM less than ten micrometers in aerodynamic diameter (PM_{10})[1]. The mass basis for the 24-hour standard was amended from 260 $\mu g/m^3$ to 150 $\mu g/m^3$, which matches this new size range of PM based on collected data analyzed during the standard determination. The annual standard concentration was set to 50 $\mu g/m^3$. The secondary standard is identical to the primary standard.

Section 109(d) of the CAA requires EPA to periodically review the NAAQS. Results from research has shown that particulate matter smaller than PM_{10} is important for health effects. In 1997 [FR, 1997], the EPA Administrator made the following revisions to the NAAQS for PM [EPA, 1999]:

---

[1] Aerodynamic diameter is defined as the diameter of a spherical particle with unit density (density of water) that will settle in quiescent air at the same rate as the particle in question.

- 24-hour and annual primary standards were added for PM with an aerodynamic diameter of 2.5 micrometers or less ($PM_{2.5}$). (Note: Upper 50% cut-point [2] of 2.5 micrometers.)
- The 24-hour standard is met when the 3-year average of the 98th percentile of the 24-hour concentrations is less than or equal to 65 $\mu g/m^3$.
- The annual standard is met when the 3-year average of the annual arithmetic mean is less than 15 $\mu g/m^3$.
- The PM_{10} 24-hour standard was retained but revised to be based on the 3-year average of the 99th percentile of the 24-hour concentrations.
- The PM_{10} annual standard was left at 50 $\mu g/m^3$, but based on a 3-year average of the annual arithmetic mean.

These new criteria are related only to the very small PM, which are common to combustion sources. Figure 1 displays typical sizes of various PM [Owen, 1992] while Figure 2 shows the large contribution from fuel combustion to the smaller particle concentration, typically less than PM_{10}. Also of note is the off-highway category, of which airport emissions are a significant percentage. Emissions are also increasing due to a growing demand for commercial air traffic. Figure 3 shows the projected increase in U.S. airline passengers [EPA, 1998].

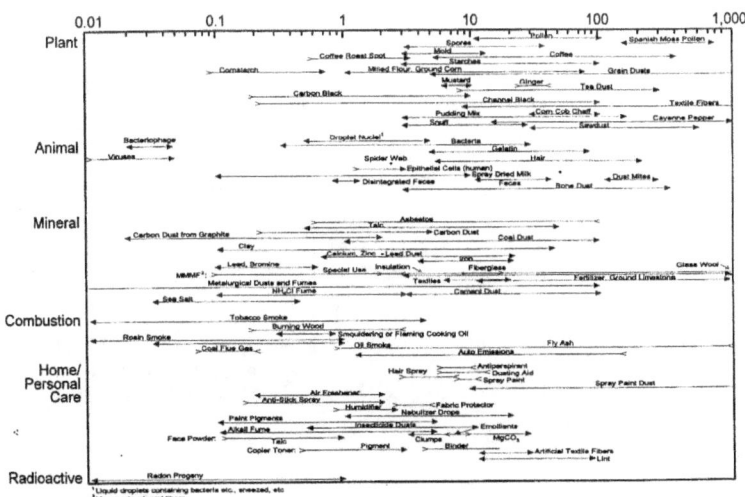

Figure 1. Sizes of Various Types of Indoor Particles

---

[2] The 50% cut-point is the limit of the size range where one-half of the particle size in question is effectively captured. PM with a greater size is collected by more than one-half whereas PM with a smaller diameter is not captured with 50% efficiency.

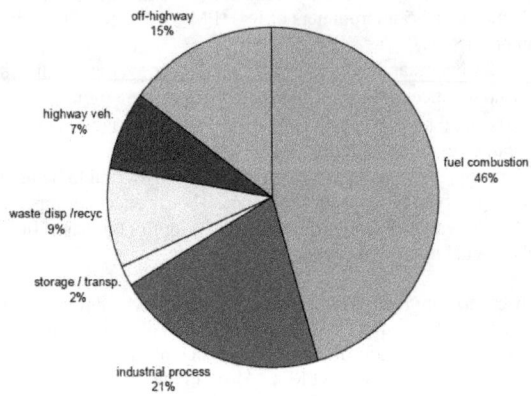

Figure 2. Nationwide Comparison of Estimated PM_{10} Emissions

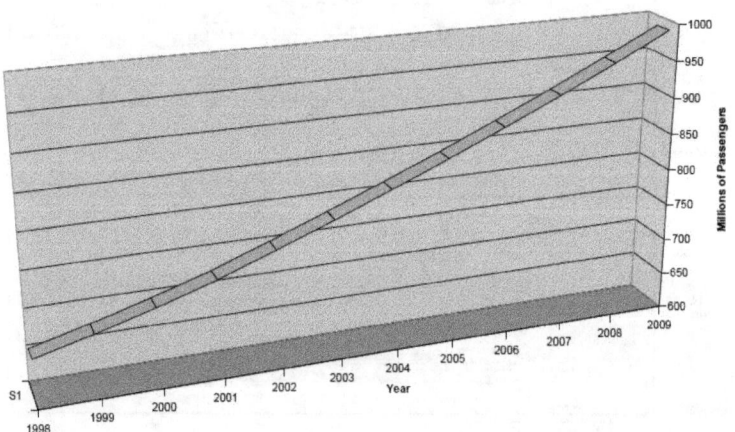

Figure 3. Projected Increase in U.S. Airline Passengers

10

2.2 Aircraft Particulate Matter Research

Results have been published by various groups addressing specific research on the characteristics of PM from aircraft sources. In order to present these results in a meaningful way, each group's results are first reported and then findings from all groups are summarized.

2.2.1 Society of Automotive Engineers (SAE)

The SAE has developed methodologies for many types of measurements. Working through international technical committees, effective standardization of measurement procedures is commonly promulgated as recommended practices. Often, these practices become official international standards. In the late 1960s, it was recognized that an approach for measuring PM from aircraft would be needed. The immediate need was the mitigation of visible plumes. Because of measurement limitations and the desire to eliminate visible plumes, a methodology based on the Ringelmann Smoke Number was devised and the methodology standardized [SAE, 1970]. This methodology is designated as the Aerospace Research Procedure (ARP) 1179 and has been updated twice since its original release. Revision B, the latest, was released in 1991. The methodology involves collection of particles on a porous filter medium (Whatman Number 40 filter) and measuring the light reflectance from the filter. However, since the primary collection mechanism is impaction, [3] most of the smaller particles are not captured. Accordingly, the smoke number does lend useful information about the visible plume behind an aircraft, but is problematic to use for any mass prediction technique or health-related analysis.

The smoke number methodology is still in use today and is the recommended test procedure for transport-category aircraft. The database is maintained by the Defense and Environmental Research Agency (DERA) in the United Kingdom for the International Civil Aviation Organization (ICAO). The aircraft-specific smoke number did fulfill its planned purpose, as the plumes from most modern commercial aircraft are primarily invisible.

The SAE's E-31 Committee is involved with trying to determine a more appropriate methodology for measuring PM transport-category aircraft.

2.2.2 United States Department of Transportation (U.S. DOT)

Initial research was reported in 1971 by the U.S. DOT Transportation Systems Center (TSC), now the Volpe Center, in Cambridge, Massachusetts [Broderick, 1971]. The reported results were very limited for PM although it was recognized that the size range for most PM was below 1 micrometer. An important question was addressed in this report. "Since particulate emissions are generally measured by weighing filters and the mass of a particle is proportional to the cube of its diameter, a question arises as to

---

[3] Impaction occurs when a particle has sufficient mass to break away from the drag force of air going around an object and hits or impacts the object.

whether "smokeless" engines merely produce greater quantities of small particles instead of fewer, larger particles." This is of particular importance, considering the recent findings on very small PM and related health effects.

Realizing that the usefulness of the smoke number may be limited and does not allow for accurate mass-based predictions, work continued at TSC as part of the Climatic Impact Assessment Program (CIAP) [Broderick, 1972]. Actual testing took place at Arnold Air Force Base near Tullahoma, Tennessee. A YJ93-GE-3 afterburning turbojet was studied. This engine is used in the XB-70 supersonic aircraft. The test facility personnel controlled the engine's inlet temperatures and pressures, effectively allowing at-altitude simulation of the aircraft. Most testing was for engine settings in the supersonic region, but some tests were also conducted without afterburner and in a military engine setting, which would be more typical of results at subsonic speeds. Instrumentation consisted of a point-to-point electrostatic precipitator designed to capture samples for analysis, using a transmission electron microscope. The sample probe used was placed very close (reported as a few inches) to the exhaust exit plane of the engine.

Data in the form of electron microscope images were presented for two of the seven test conditions. From these images of primarily the non-volatile PM, it was concluded that the structure of the emitted particles was found to be very irregular and variable in size, primarily smaller than a micrometer, with no immediate similarities between different particles. The number, structure and size also appeared to change with test conditions. The report recommended that any future research be focused on particles smaller than 0.5 micrometers. Follow-up testing was planned involving a J85-GE-5 engine. The results of these tests have currently not been published

2.2.3 Battelle Corporation

The Battelle Corporation's Columbus Division, in cooperation with the U.S. Air Force Engineering and Services Center at Tyndall Air Force Base, conducted aircraft emission characterization research [Spicer, 1988]. The engines included: TF33-P3, TF33-P7, and J79 (with a smokeless designation). Testing was performed using JP-4 fuel at throttle settings of idle, 30 percent, 75 percent, and 100 percent. Multiple PM measurement techniques were included: (1) Teflon coated, (2) glass fiber particulate filter, smoke number, (3) diffusion Battery Condensation Nucleus Counter, and (4) Electrostatic Particle Sampler. The output of these measurements included the mass amounts from the filter analysis, the smoke number, and particle count (including aerosols). Table 1 lists the important findings of the measurements.

Table 1 shows that the JP-79-17G was not only designed to be "smokeless," but does indeed seem to emit less mass as well. Also of note is that the mass concentration for the TF33-P3 is greater at all power settings except idle when compared to the TF33-P7 engine. This tends to indicate that the new combustor designs and the relationship with fuel flow through the combustor has resulted in less mass being emitted in these engines. Another important conclusion stated in the report is that "...there are relatively small numbers of particles of diameter greater than 0.24 μm." This was also thought to be the

Table 1. Reported Engine Measurements [Spicer, 1988]

ENGINE	POWER SETTING (%)	SMOKE NUMBER	EXHAUST VOLUME (m^3)	CONCENTRATION (mg/m^3)	PARTICLE NUMBER (1000/cc)
JP-79-17G	Idle	20.3	0.51	1.71	1320
	30	24.1	1.41	0.60	353
	75	16.1	1.48	1.45	1870
	100	22.6	1.31	4.22	1700
TF33-P3	Idle	20.4	0.61	6.27	5230
	30	36.0	0.35	16.6	5500
	75	54.0	0.57	32.0	7750
	100	59.4	0.61	36.2	4530
TF33-P7	Idle	20.0	0.28	7.39	6250
	30	35.3	0.40	11.6	7179
	75	51.6	0.55	24.6	6364
	100	52.5	0.78	20.8	3608

reason that smoke number increased with increasing throttle settings while the particle concentration did not increase in a consistent fashion. Light reflectance was significantly affected by particle sizes less than 0.2 micrometers was thought to be one reason non-linear behavior is observed when mass emissions and smoke number are compared. The smoke number was also affected by larger particles being emitted at the higher throttle settings.

2.2.4 Intergovernmental Panel on Climate Change (IPCC)

The IPCC has published a special report characterizing global emissions from aircraft at altitude [IPCC, 1999]. This highly regarded reference includes a consensus of information on the PM emissions from aircraft. An emphasis is placed on aerosols in the report. As stated earlier, air-borne PM is considered and atmospheric aerosol.

The IPCC report states emissions from aircraft engines include soot, metal particles, liquid aerosols such as water vapor, oxidized sulfur in various forms, chemi-ions (charged molecules), nitrogen compounds, and unburned hydrocarbons. A large number of these components are emitted (10^{17}/kg fuel burned) and are in the size range of 1 to 10 nanometers. These particles may be volatile (prone to evaporate) and can form sulfuric acid, chemi-ions, and water vapor. In addition, they can grow in size while in the aircraft exhaust plume through coagulation, uptake of water vapor, and by condensation. For example, in a "young" plume, the conversion of fuel sulfur to sulfuric acid is likely to be in the range of 0.4 to 20 percent [IPCC, 1999]. These emissions have led to increased concentrations of aircraft-produced aerosols near air traffic corridors. Aerosol emissions may also lead to contrails (estimated to be 0.5 percent of the sky over central Europe) and increased cloud cover. Consequently this invisible trail of aerosols left in the aircraft path

13

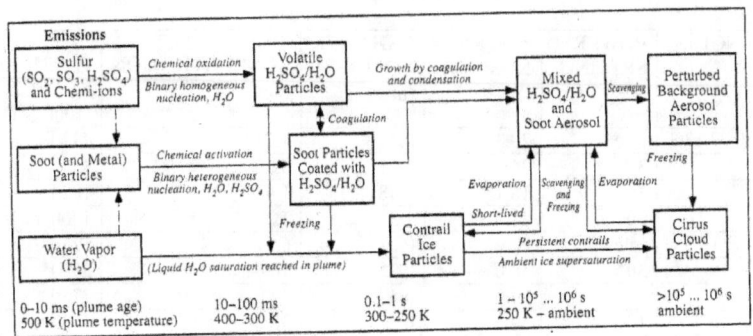

Emissions						

Figure 4. Aerosol and Contrail Formation Processes in an Aircraft Plume and Wake as a Function of Plume Age and Temperature [IPCC, 1999]

may affect the global climate. Figure 4 provides an overview of the aerosol emission and formation process.

The volatile particles in the aircraft exhaust form as a result of nucleation processes from the precursor aerosols[4]. Some typical aerosols and their characteristics are presented in Table 2 for reference. To help visualize these size ranges, Figure 5 has also been included. It shows the relative size distributions of the various aerosols immediately behind the aircraft.

Directly emitted (primary PM emission) is solid soot particles. Soot is composed of all primary, carbon-containing products resulting from the incomplete combustion processes in the engine, namely the optically black carbon fraction and the nonvolatile (gray) organic compounds. Measurement of the soot particles is required during testing of transport-category aircraft using the smoke number. However, this approach is considered somewhat unreliable, since the primary soot particles which are captured as a part of the testing process are almost all greater than 100 micrometers in diameter resulting in the smoke number determination being dominated by the larger particles.

The soot particle numbers for a variety of contemporary engines show a scatter of around 10^{15}/kg of fuel burned[5]. This results in the soot particles being about 100 times less prevalent than the volatile aerosol particles on a number basis. There would seem to be

---

[4] Nucleation is the process of a small particle acting as a collection point that allows the particle to grow in size.
[5] In this nomenclature, it is implied that the units are the number of particles emitted per kilogram of fuel burned.

Table 2. Summary of Number Mean Radius, Number Density, and Surface Area Density for Sulfate and Soot Particles in Aircraft Plumes, and in the Background Atmosphere, and for Ice Particles in Contrails and Cirrus [IPCC, 1999]

	Radius (μm)	Number Density (cm⁻³)	Surface Area Density (μm² cm⁻³)
Sulfate			
Plume (1 s)[a]	0.002	$(1-2) \times 10^7$	500–1000
Background (10–12 km)[b]	0.01–0.1	50–1000	1–6 (10–40)
Background (20 km, non-volcanic)[c]	0.07	5–10	0.5–1
Background (20 km, volcanic)[d]	0.2–0.5	10–100	10–40
Soot			
Plume (1 s)[e]	0.01–0.03	$5 \times 10^4 - 5 \times 10^5$	50–5000
Background (10–20 km)[f]	0.05–0.1	0.01–0.1	$3 \times 10^{-5} - 3 \times 10^{-3}$
Ice			
Young Contrail (0.1–0.5 s)[g]	0.3–1	$10^4 - 10^5$	$10^4 - 10^5$
Persistent Contrail (10 min to 1 h)[h]	1–15	10–500	$10^3 - 10^3$
Young Cirrus[i]	5–10	1	$10^2 - 10^4$
1992 Aircraft Perturbation			
Sulfate Aerosol (10–12 km, 50–60°N)	0.01	90–900	0.1–1.1
Soot (10–12 km, 50–60°N)	0.02	3–30	0.02–0.2

[a] Detectable only by ultrafine particle counters (particles smaller than 2–3 nm radius are not detected). Calculations by Yu and Turco (1997) for average FSC consistent with observed data.
[b] Properties highly variable; size distributions often bimodal. Ranges include small (> 10 nm) particles. Large particle mode (~100 nm) often similar to mann, 1993; Yue et al., 1994; Schröder and Ström, 1997; Solomon et al., 1997; Hofmann et al., 1998. High range of tion data and represents mixtures of aerosols and subvisible clouds.
[c] Bormann et al., 1997; Thomason et al., 1997b.
[d] Petzold et al., 1999.
[e] gest atmospheric lifetimes are measured by wire impactors (Sheridan et al., 1994; Blake and Kato, 1995; Pueschel et al., ... area introduced by fractal geometry of particles. ... ld et al., 1997.
[f] Values representative of contrail core for low ice-supersaturation (Heymsfield et al., 1998a; Schröder et al., 1998b) (see also Sections 3.4.4 and 3.6.3). For large particles are observed for large ice-supersaturation (Knollenberg, 1972; Gayet et al., 1996).
[g] Ström et al., 1997; Schröder et al., 1998b. Larger values are observed in warm cirrus clouds (Heymsfield, 1993; see also Sections 3.4.4 and 3.6.3).
[h] Results of fuel tracer simulations discussed in Section 3.3.4. Values shown represent upper bounds to zonal mean perturbations caused by emissions of the 1992 aircraft fleet. Results are representative of flight levels at northern mid-latitudes and are calculated using the range of values of computed tracer concentrations from all models and assuming a fuel sulfur content of 0.4 g/kg fuel, a 5% conversion of sulfur to sulfate aerosol, an EI(soot) of 0.04 g/kg fuel, and a mean particle size of 10(20) nm for sulfate (soot) particles.

no dependence between the sulfur content of the fuel and the soot emitted. In terms of mass, this number of particles (10^{14} to 10^{15}/kg of fuel burned) would result in about 0.01 to 0.2 g/kg of fuel burned, on average. The authors wish to point out that the fuel burn rate is highly dependent on aircraft/engine combinations. For example, the older Concorde/Olympus and T-38 engines show exceptionally high emissions as compared to the newer engine designs. In fact, some modern aircraft engines have reduced the number of soot particles emitted to about 10^{13}/kg of fuel burned.

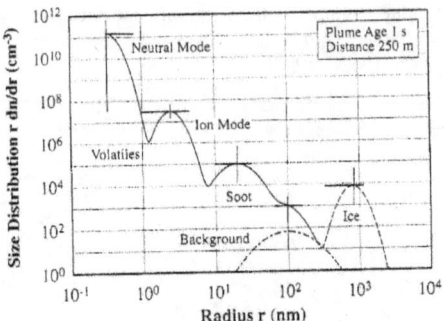

Figure 5. Size Distribution of Various Aerosol Types Present In Young Jet Aircraft
Exhaust Plumes [IPCC, 1999]

The soot particles are reported in the IPCC report to be primarily composed of individual,
nearly spherical particles called spherules. The mean radius of these particles is typically
in the 10 to 30-nanometer range. This exceeds the size of the volatile aerosol particles in
a young plume (see Table 2 and Figure 5). It has been observed in this study and others
that soot particles can aggregate and form a complex chain structure with time. The size
of the particles also depends on aircraft altitude. The smallest soot particles, at cruise
conditions, are rapidly immersed in the background aerosol droplets, undergo
coagulation, and as such only larger radii particles of about 50 to 100 nanometers are
observed. The reported soot surface area at the engine exit is in the range of 5,000 to 10^5
microgram-square-meter per cubic centimeter. Also, although the sulfur content of the
fuel does not correlate well to the soot particle emissions, sulfur may be incorporated into
the emissions as part of the soluble mass fractions, or found on the surface of the
particles. The irregular surface features of the PM emitted is conducive to chemical
reactions can also increase chemical reactivity and amplify the heterogeneous nucleation
processes.

Metal particles directly emitted by aircraft engines (primary PM) include fragments due
to engine erosion and metals that are in the fuel but not oxidized during the combustion
process. Elements such as Al, Ti, Cr, Fe, Ni, and Ba are estimated to be present in the
parts-per-billion by volume range at nozzle exit planes. Corresponding concentrations
are less than that of soot and are in the range of 10^7 to 10^8 /kg of fuel burned.

2.2.5 International Civil Aviation Organization (ICAO)

The International Civil Aviation Organization (ICAO) maintains a database at the
Defense and Environmental Research Agency (DERA) in the United Kingdom. It is the
only comprehensive database in existence and includes fuel flow rates and emission

indexes for common air pollutants from aircraft engine exhaust. These pollutants include carbon monoxide, nitrogen oxides, sulfur oxides, hydrocarbons, carbon dioxide and smoke number for most transport-category aircraft. The data are reported for various throttle settings that represent common modes of aircraft operation. Table 3 shows the throttle settings and the related mode for the aircraft.

Table 3. Aircraft Throttle Settings and Related Aircraft Modes in the ICAO Database

Throttle Setting (% of Full Throttle)	Related Aircraft Mode
7	Taxi/idle
30	Approach
85	Climbout
100	Takeoff

Unfortunately, PM is represented by the Smoke Number (SN) in the ICAO database. It has been shown by multiple researchers that the SN and the mass of PM emissions are not well correlated and will be discussed in this report. This is because SN is a measure of light reflectance on a filter that has captured PM from a small portion of the aircraft exhaust. As observed, the filter that is used is a Whatman Number 40 with openings in the fiber on the order of 200 nanometers while the great majority of PM in aircraft exhaust is an order of magnitude less in size. Consequently, only the larger PM is captured and the SN is valid only for this larger PM -- and then, only in terms of light reflectance.

2.2.6 Deutsches Zentrum fur Luft-und Raumfahrt (DLR)

In 1998, Petzold and Dopelheuer [Petzold, 1998] published emission indexes for black carbon in the range of 0.01 to 0.5 g per kg of fuel burned. The number of particles was reported to be as high as 10^{15} particles per kg of fuel burn. The report made the point that emission indexes have to be adjusted for fuel flow, and concentration of PM is related to the temperature and pressure of the exhaust gas. Of note in this work was that the black carbon emission index was a linear function of thrust, and hence fuel flow.

Work reported on by Dopelheuer the next year implied that correlation with the SN and mass emitted may be possible [Dopelheuer, 1999]. He did note that soot production and oxidation were very complex, and not well understood. Further, the study was complicated by the non-homogeneous flow and temperature fields in the combustion chamber, the influences of the injection system and combustor technology, the fuel qualities, and a lack of measured data. Although the ICAO database contains only smoke number for particulate matter, it is fairly comprehensive in terms of aircraft engines and is widely accepted; therefore it becomes the starting point for many approximations of emission indexes. Unfortunately, when compared to the SLS-thrust (Sea Level Static) in percent, the trends in mass emissions of PM are quite different for various engine types and even similar engines with different combustor designs. As a first cut to develop a semi-empirical method, results from Whyte [Whyte, 1982], Hurley [Hurley, 1993], and Champagne [Champagne, 1971] were plotted by DLR and analyzed. Figure 6 is reproduced from the latter paper. It shows that the concentration of soot (in milligrams per cubic meter of exhaust gas) may be correlated to the SN. Also shown is the statistical fit through the data. Although this is a first cut approximation, DLR suggests that the close agreement of the curves by Champagne and Hurley could be used for prediction of concentrations up to 6 mg/m^3 and the work of Whyte applied above that cut point. A necessary reference function, unexplained in the text, was then explored based on the relationship of the soot concentration to the Combustor Inlet Temperature (T_3) at SLS. These results are shown in Figure 7. It can be seen in this Figure that the functions vary widely for the two engines analyzed, with the concentration of soot being much lower in the lower power setting for the more modern CFM56-5C2 engine. This was thought to be more likely to the differences in injector systems and combustor technologies of the more modern engines when compared to the older engines. This difference would seem to point out that individual coefficients would need to be determined for each engine type and possibly each engine/airframe combination.

The DLR report also outlines that, using the SLS reference function, other than sea level emission indexes may be predicted by Equation 1:

$$Csoot = Csoot, ref \, [\Phi / \Phi_{ref}]^{2.5} \, [p_3 / p_{3, ref}]^{1.35} \, [(exp(-20000/T_{fl})) \, /(exp(-20000/T_{fl, ref}))] \quad [1]$$

Where: Csoot = concentration of soot in mg /m^3
           Φ = equivalence ratio
           p_3 = combustor inlet pressure
           T_{fl} = flame temperature

The reference values are at the same combustor inlet temperature at SLS conditions.

Although the associated developmental methodology was not provided, the paper did cite an average emission factor of 0.008 g/kg for an A300 aircraft equipped with a CFM56-5C2 engine. When an 8000-kilometer trip was considered, cruise soot emissions were much greater than all other modes. When predicted near an airport (take-off and descent modes of operation) the value would be 0.03kg. When aircraft technology was further investigated it was found that the A300 aircraft equipped with a CF6-50C2 engine and a B767 with a CF6-80C2 engine performed similarly over a 4000 km trip with the B767 emitting more soot. However, when compared to the older B707 with a JT3D-3B engine, the B707 was an order of magnitude higher in soot emissions than the other two aircraft. Results of engine testing performed for the E3E program at the Stuttgart University Altitude Test Facility, was published by Dopelheuer and Wahl [Dopelheuer, 2000]. The paper highlighted the main findings that had previously been presented in poster form. The paper reported characterization of aircraft/engine-generated soot, using a Scanning Mobility Particle Sizer (SMPS) and a Scanning Electron Microscope (SEM). This testing was performed using a Rolls Royce Deutschland core engine with a low emission annular combustor. Emissions that would typically occur during takeoff, climb, cruise, descent, and taxi were analyzed. Table 4 summarizes the reported results. In addition to these

18

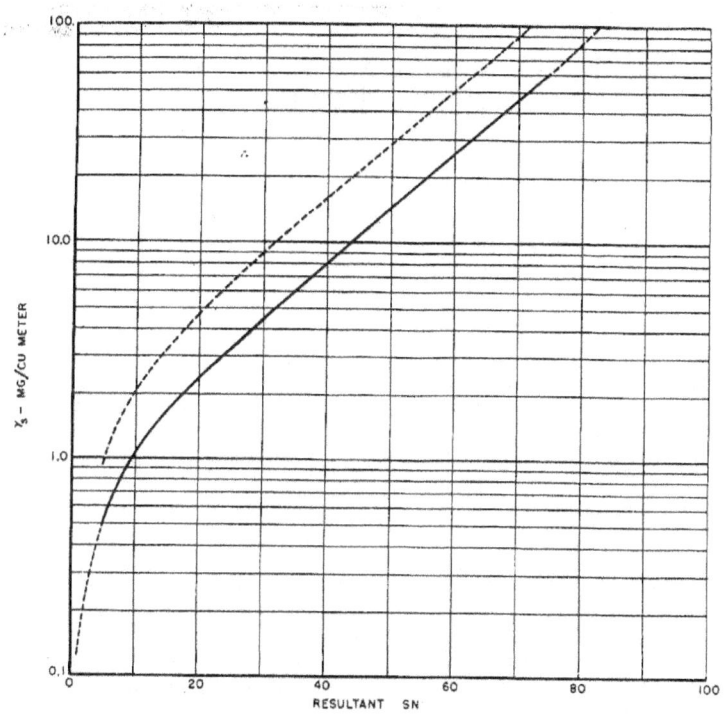

Figure 6. Relationship Between Resultant SN and True Smoke Density [Champagne, 1971]

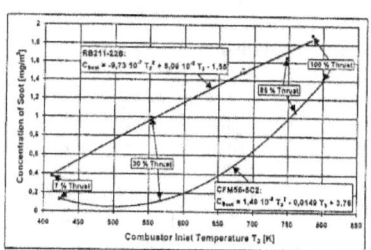

Figure 7. Concentration of Soot Versus Combustor Inlet Temperatures for CFM56-5C2 and RB211-22B at SLS [Doppelhauer, 1999]

19

Table 4. Summarized Results Reported by DLR

Aircraft Mode	Density (x 10 /cm^3)	Mean Size (nm)	Approximate Range (nm)	$EI_i/EI_{takeoff}$	Comments
Take-off	2.09	36.4	6 – 200	1.0	Bi-modal dist
Climb	1.73	36.0	6 – 225	0.58	
Cruise	1.66	31.8	6 – 200	0.25	
Descent	1.52	29.3	6 – 175	0.18	
Taxi	0.17	21.8	6 – 100	0.15	

results it was reported that:

- the PM structure is of a fractal nature (irregular in shape);
- a log-normal distribution is appropriate to characterize the soot aerosol;
- the take-off mode contains a second, smaller diameter size mode distribution of PM;
- the number concentration and particle diameter decrease with decreasing power; and,
- the standard deviation shows a slight decrease with decreasing power.

Of secondary interest is a global emissions inventory that was also performed by DLR, based on 34 aircraft engine combinations and in-house flight performance models. The maximum PM concentration values occur in North America and Europe, as expected, and were reported to be 76 g/km^2 per month. The global mass emissions were estimated to be 330 tons per month, with the average emission factor of 0.038 grams of soot per kilogram of fuel burned. These estimates were based on the derived emission indexes, the number of flights, and types of aircraft used.

Because the ICAO database is generally complete and internationally accepted, DLR also examined the correlation between the measured PM parameters and the ICAO-reported SN. Using the SEM it was shown by Dopplehauer that the Whatmann No. 4 filter used in the SN testing has large spacing between the fibers in comparison to the particle sizes measured from aircraft engine exhausts. As such, capture of only the larger particles would be expected. Since the vast majority of particles would not be captured, this makes correlation difficult.

DLR has also evaluated aircraft at altitude involving a chase plane [Petzold, 1999]. Using the DLR Falcon research aircraft, measurements of carbon black have been performed for three aircraft/engine combinations: an A310-300 aircraft equipped with a CF6-80C2A2 engine; ATTAS[6] aircraft / Rolls-Royce/Snecma M45H Mk501 turbofan engine; and a B737-300 aircraft / CFM56-3B1 engine. Data reported were primarily taken during the 1998 SULFUR6 experiments. Measurements were made with the chase plane at distances less than 500 meters at altitudes from 8.3 to 10.7 kilometers.

---

[6] This is a research aircraft used by the German Research Lab, DLR.

Equipment included: passive-cavity aerosol spectrometer probe, forward scattering spectrometer probe, condensation nuclei counter in various modes, particle soot absorption photometer, integrating nephelometer, and optical probes. A comparison between the ATTAS and B737-300 aircraft showed some interesting trends. For the newer engine of the B737-300 (CFM56-3B1), the mean particle size was smaller and the range of the overall particle size distribution was not as large as for the ATTAS aircraft / M45H Mk-501 engine. In addition, the calculated mass emission index was significantly less for the B737-300. Table 5 shows the reported data for these tests. It is important to note the continuing trend of smaller mass amounts and smaller particles with more recently manufactured engine designs. In fact, at altitude the new engines emitted almost an order of magnitude less mass. When all aircraft were analyzed, including the A310, it was determined that "...the number emission indexes of modern engines are $< 10^{15}$ kg^{-1}." Also of note was, that for the CFM56-3B1 engine the primary black carbon-mode-count median diameter was 0.025 micrometers; and for this engine PM had a coagulated count median diameter of 0.15 micrometers. This implies to the authors that the emissions have a bi-modal distribution. Other researchers [Yu, 1998] have used mono-modal lognormal

Table 5. Calculated BC Mass and Particle Emission Indices for Common Airframe – Engine Combinations for Whole Flight Missions and for 100% Thrust at Sea-Level-Static Conditions [Petzold, 1999]

Airframe	Engine	Engine Certification Year	Engine EI(BC) at 100% SLS Thrust [*]	Flight Mission EI(BC) [*]	Flight Mission EI(N) [*]
B707	JT3D-3B	1962	0.829	0.294	91.2
L1011	RB211-22B	1971	0.089	0.038	11.8
L1011-250	RB211-524B	1975	0.1	0.027	8.8
A300	CF6-50C2	1977	0.02	0.012	3.7
DC 10-30	CF6-50C2	1977	0.02	0.01	3.1
B747-300	CF6-50E2	1977	0.02	0.009	2.8
B747-400	CF6-80C2B1F	1982	0.037	0.016	5.0
B737	CFM56-3B2	1983	0.025	0.013	4.0
MD11	PW 4460	1986	0.043	0.026	8.1
	CFM56-5C2	1990	0.052		

SLS, sea-level-static.
[*] Units are gram of black carbon per kilogram of consumed fuel.
[*] Units are 10^{14} particles per kilogram of consumed fuel.

distributions with a mass mean diameter of 0.06 micrometers, a value in the middle of the DLR observations. Figure 8 shows the differences between a single primary mode and a bi-modal distribution where particle size distributions tend to fall into a single or dual distribution. An observation from the paper was, that for the ATTAS the cruise carbon black emission index was roughly equal to the 30% thrust level at sea level conditions. This indicates that less than one-third of the mass is emitted at altitude as compared to ground level operations.

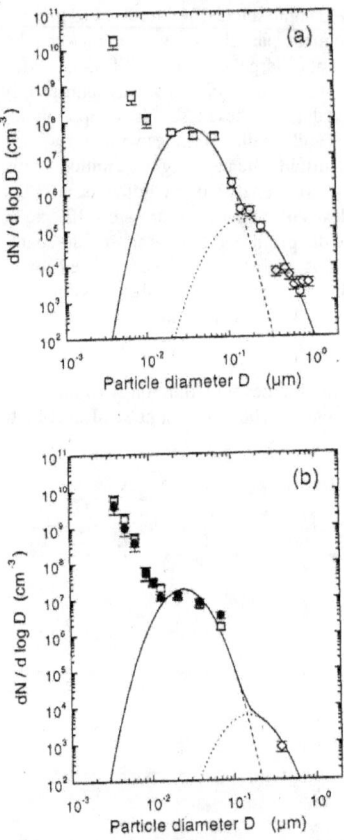

Figure 8. Size Distributions of Exhaust Aerosol in the Plume of the (a) ATTAS and (b) B737 [Petzold, 1999]

2.2.7 National Aeronautics and Space Administration (NASA)

NASA responsibilities have been expanded to include the development of advanced combustor technologies for both subsonic and supersonic aircraft. In order to address these expanded responsibilities, NASA identified the need for a database to help assess the atmospheric impact of aircraft emissions. The development of this database is being performed under the Atmospheric Effects of Aviation Project (AEAP). Primary participants in this work include the University of Missouri at Rolla and Arnold Engineering Development Center (Arnold Air Force Base). The roles of these organizations are discussed in more detail in the following sections.

2.2.7.1 University of Missouri at Rolla (UMR).

UMR is working with NASA to develop a measurement methodology that provides reliable quantification of PM. The technique, referred to as UMR-MASS (University of Missouri-Rolla Mobile Aerosol Sampling System), is being carefully defined and has previously been used in multiple evaluations of aircraft PM, using various configurations of the components[7]. Extractive samples can be used to measure the condensation nuclei for all aerosols on a real-time basis. A laser aerosol spectrometer is used for real time measurement of larger aerosols (1.0 to 30 micrometers). Detailed characterization of radial profiles of static temperature and pressure, and radial/axial measurements of flow velocities are performed during this testing as well. A needle-to-grid electrostatic precipitator is used for aerosol collection and deposition on electron microscope grids. Aerosol samples captured in prepared tanks can be used to measure size distribution, growth, and/or hydration using electrostatic aerosol classifiers.

In addition to this development, UMR is also preparing a PM literature review for NASA. However, this will be focused on engine design and measurements and is likely to be of little use for the current mass approximation methodology needed by the FAA to meet Federal regulations.

As part of the NASA FIRE project (First ISSCP - International Satellite Cloud Climatology Project) Regional Experiment), field sampling of jet exhaust aerosols was accomplished by UMR [Hagen, 1992]. Using the National Center for Atmospheric Research (NCAR) Sabreliner aircraft, the aerosols in aged plumes were measured at altitude using two electrostatic aerosol classifiers. The engines tested were Pratt and Whitney PT 6-42, JT 12A-8, and JT 15-D-4. It was determined that the PM size for the raw (unchanged) exhaust was between 0.014 and 0.2 micrometers, peaking in the 0.03 to 0.05 micrometer range. Of particular interest was that the aircraft plumes could be detected with the aerosol sampling methods for time intervals of 15 to 20 minutes after the aircraft had passed.

Using the UMR-MASS instrumentation system, a bimodal PM distribution was observed in the jet exhaust at altitude [Hagen, 1996]. It was found that at about 8 kilometers behind the aircraft at cruise conditions, the distributions were centered around 50 nanometers and 0.1 to 0.2 micrometers. Neither of the distributions displayed the sharp drop-off in particle concentration at the small-particle end of the distribution that had been previously found in ground-based, engine test stand studies. Also, for the test stand studies, a relatively low soluble mass fraction was observed (usually below 0.2) meaning

---

[7] A partial listing of UMR research partners include DLR, the National Climatic Aero Research (NCAR), NASA Lewis, Pratt and Whitney, Arnold Engineering Development Center, McDonnell Douglas Aerospace – East (MDAE), and others.

that most particles were solid. But at altitude, engines were found to produce particles having high soluble mass fractions. This tended to indicate that the small particles might be formed by binary nucleation of sulfuric acid aerosols or by the heterogeneous nucleation on ion clusters.

In 1998, the UMR group published another article that further refined its work using the UMR-MASS system [Hagen, 1998]. The primary analysis centered on a Rolls Royce RB211 engine in a B757 airframe. The measurements were done as part of the NASA project, SUbsonic aircraft: Contrail & Clouds Effects Special Study (SUCCESS). The equipment was in the NASA DC-8 research aircraft. This study showed that a difference existed between measurements performed on the ground and for airborne aircraft, with the particle population being lower on the ground. It also showed that burning high sulfur fuel (700 ppm) tended to increase the pollutant emission indexes as compared with the low sulfur fuel (70 ppm). Figure 9 shows these results for the various cases while Table 6 shows the change of all emission indexes with the change in fuel sulfur. The mass indexes in Table 6 vary by an order of magnitude. Sulfur conversion efficiencies to PM were found to be 26 percent for the high sulfur fuel and 19 percent for the low sulfur fuel. It should be noted that the mass emission indexes were derived assuming a particle density of 1.9 grams per cubic centimeter. In many other similar analyses, a value of 2.0 grams per cubic centimeter had been used.

Table 6. Emission Indices [Hagen, 1998]

	High Sulfur	Low Sulfur	Units
EI(numb,tot)	$(2.6\pm0.4)E+15$	$(2.8\pm0.3)E+14$	Number/kg-fuel
EI(numb,nv)	$(1.2\pm0.2)E+14$	$(7.0\pm0.3)E+13$	"
EI(area,tot)	$(3.6\pm0.6)E+4$	$(3.9\pm0.4)E+3$	cm^2/kg-fuel
EI(area,nv)	$(1.6\pm0.3)E+3$	980 ± 40	"
EI(mass,tot)	0.62 ± 0.10	0.067 ± 0.007	g/kg-fuel
EI(mass,nv)	0.042 ± 0.007	0.026 ± 0.001	"

(errors quoted are 1σ)

nv = non-volatile

24

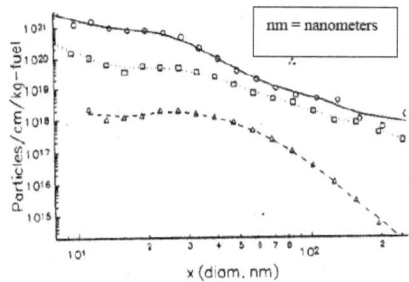

Figure 9. Differential Size Distributions in Emission Index Format (number of particles per unit size interval per kg – fuel burned0 for Two Plume Encounters; May, 1999, B757
Squares – Low Sulfur Fuel
Circles – High Sulfur Fuel
Triangles – One Engine During Ground Test, Same Aircraft
[Hagen, 1998]

2.2.7.2 Arnold Engineering Development Center (AEDC)

Like the UMR group, AEDC is also involved in NASA research. As part of the NASA Atmospheric Effect Of Aviation Project (AEAP), AEDC participated in controlled jet engine tests to simulate sea-level-static (STS) test data, as well as data taken at altitudes up to 15 kilometers [Howard, 1996]. Altitude testing was primarily done at 0.8 Mach to simulate commercial aircraft cruise operations. The engine tested contained an annular combustor and was typical of a 133 to 178 kN (30-40,000 lbf) thrust commercial engine. For this work, both extractive probes (sample taken with a rake probe, 12 centimeters downstream from the nozzle exit plane) and non-intrusive optical techniques were used. Figure 10 shows the equipment setup. Use of a smoke meter consistent with SAE methodologies provided the smoke number for three tests. Engine aerosol emissions were characterized using the UMR-MASS.

The study reported that PM from the tested engine had a mean aerosol number-based emission index from all measurements of $2.2 +/- 0.7 \times 10^{13}$, while the mass-based emission index was 0.012 +/- 0.001. These numbers were reported to be small compared with other measurements previously reported using UMR-MASS. Smoke numbers reported were also small (–1.1, 0.7, and 1.3) and it was noted that the low smoke numbers correlated with these low particle emission indexes. The typical size distribution was log-normal (also seen in previous testing using the UMR-MASS) with a peak in the vicinity of 20 to 40 nanometers. Table 7 shows the emission indexes as a function of test parameters and Figure 11 shows a plot of the particle size distribution.

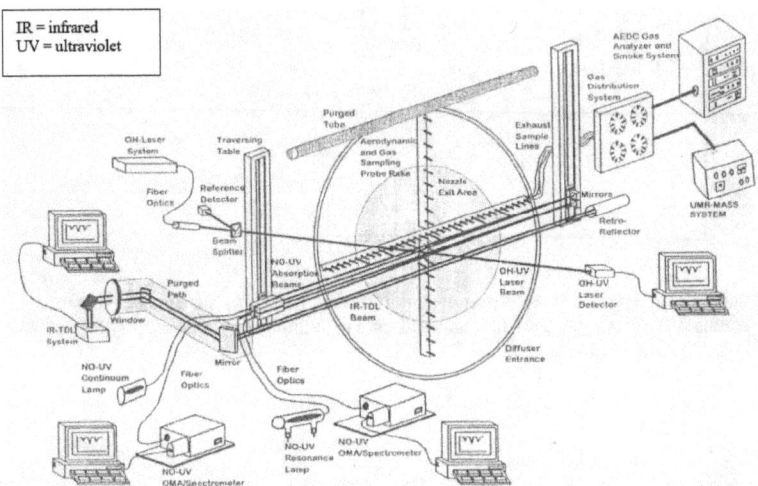

Figure 10. Turbine Engine Exhaust Emissions Measurement Illustration [Howard, 1996]

It was also reported that the volatile component of the PM is small, and could be considered negligible. The particle emission indexes were not determined to be strongly dependent on thrust, altitude, combustor inlet pressure or inlet temperature. An inverse relationship between combustor efficiency and the particle emission index was shown, but data were limited. The researchers were concerned about the measurement techniques adequately describing the PM. As with other tests, the nature of the aerosol emission size distribution was shown to be log-normal with sizes in the 20 to 40 nanometer range. The particles displayed properties of being fractal in shape and tended to aggregate. The particle morphology was reported to need further study to determine what happens later in the exhaust plume.

3.0 APPROXIMATION METHODS USED IN PARTICULATE MATTER MASS ANALYSIS

 Because of the regulatory requirements for both air pollution source control and the need to comply with the NAAQS, estimates must be made of the mass of emissions. To perform this task, emission indexes are needed. The PM indexes are a function of aircraft operational conditions. For example, an emission index, often referred to as an emission factor, could be grams of PM per kilogram of fuel burned. Very few emission indexes based on measured data are available. A limited number of PM indexes are reported in

Table 7. Size Distribution Index and Associated EI's Listed as a Function of Test Parameters [Howard, 1996]

Time (h:m:s)	Time, sec	Size Distribution Run Numbers Total Aerosol	Non-Volatile	Tank Label	Fill Time, sec	Test Facility Steady-State Data Point	Sample Source (Valve Settings)	Altitude, km	Volume Fraction	EI (Number)	EI (Mass)	%CO₂
7:57:00	28620	1	2	1B(4B)	30	not listed	V+(10-14,28-33)	7.6	4.33E-11	4.04E+12	0.001494	3.18
8:22:34	30154	9	10	5A	40	28	V+(10-14,28-33)	7.6				2.72
8:39:50	31190	12	13	6A	40	30	V+(10-14,28-33)	7.6				
08:44.3	31470			2B	40	32	V+(10-14,28-33)	7.6				
8:59:30	32370			7A	40	not listed		9.1				
9:05:30	32730	21		8A	30	not listed	V+(28-33,10-14)	9.1	3.37E-10	5.93E+12	0.001736	3.10
9:21:24	33684	26	27	9A	40	37,38,40-49	V+(25-33,10-14)	9.1				
9:33:00	34360	29		10A	30	37,38,40-49	V+(28-33,10-14)	9.1				
9:34:00	34440			11A		37,38,40-49	full across	9.1	6.98E-09	7.07E+12	0.001703	3.39
10:27:20	37640	39		1C	30	37,38,40-49	full across	9.1	6.71E-11	4.59E+12	0.003620	3.42
10:29:00	37740	51	57	2C	30	37,38,40-49	H+(-1,6,19-24)	9.1	2.15E-11	4.59E+12	0.001230	3.42
10:42:08	38528	56		3B	30	37,38,40-49	single(V22)	9.1	3.07E-11	9.56E+12	0.003433	3.42
10:47:00	38820	59		4B	30	37,38,40-49	single(V4)	9.1	2.62E-09	8.13E+12	0.028174	3.58
10:54:00	39240	64		5B	30	37,38,40-49	H+(1-6)	9.1	3.30E-10	6.30E+12	0.002751	3.27
11:00:00	39600	69	77	6B	30	37,38,40-49	H+(19-24)	9.1	4.35E-11	8.82E+12	0.005589	3.26
11:02:00	39720	76		7B	30	37,38,40-49	V-(28-33)	9.1	1.15E-10	9.49E+12	0.014340	3.44
11:18:00	40680	79		8B	30	51,52	V-(28-33)	9.1	3.57E-11	7.14E+12	0.003323	3.73
11:29:00	41340	82		9B	30	54,55	V-(28-33)	9.1	4.11E-11	6.64E+09	0.000004	4.11
11:49:00	42540	87	93	10-B	30	57,58	V+(10-14)	12.2	4.15E-11	2.96E+12	0.003198	2.77
12:07:00	43620	92		11-B	40	60-64	V13,V31	12.2	1.63E-11	6.61E+12	0.002773	3.40
12:26:00	44680	95	106	1-D	40	60-64	V+(10-14,28-33)	12.2	1.44E-11	1.44E+13	0.019529	2.31
12:52:09	46320	105		2-C	50	66,67,68	V+(10-14,28-33)	12.2	4.14E-11	1.01E+13	0.004191	3.70
12:56:00	46560	108		3-C	50	66,67,68	V+(10-14,28,33)	12.2	5.49E-11	1.07E+13	0.006893	3.70
13:00:00	46800	113	119	4-C	50	66,67,68	V+(10-14,28-33)	12.2	2.09E-10	7.48E+12	0.018420	3.70
13:24:00	48240	118		5-C	50	71,72	V+(10-14,28-33)	15.2	3.44E-11	8.64E+12	0.015643	2.98
13:30:00	48600	125		6-C	50	71,72	V+(10-14,28-33)	15.2	1.43E-11	4.73E+12	0.006778	2.98
13:45:00	49500	130		7-C	90	74,75	V+(10-14,28-33)	15.2				2.14
14:00:00	50400	135		8-C	60	not listed	V+(10-14,28-33)	15.2	8.41E-13	1.52E+12	0.000910	3.18
14:08:30	50910	143		9-C	60	79	V+(10-14,28-33)	3 (SLS)				
14:18:00	51480			10-C	60	not listed	V+(10-14,28-33)	3 (SLS)	4.34E-11	3.79E+13	0.019356	3.96
14:57:00	53820	151		11-C	60	83,84	V+(10-14,28-33)	3 (SLS)	1.32E-10	1.55E+13	0.022249	2.43
15:05:00	54300	156	157	12-C	30	85-94	single(V13)	3 (SLS)	6.42E-12	1.35E+13	0.000856	3.09
15:53:00	57180	162	163	1-E	30	86,94	single(V28)	3 (SLS)				3.13
15:58:00	57480			2E	45	86,94	V+(10-14,28-33)	3 (SLS)	7.03E-11	3.11E+13	0.022435	3.27
16:07:00	58020	169		4D	45	86,94	single(V22)	3 (SLS)				3.37
16:24:00	59040	172		5D	45	96,96	single(V1)	3 (SLS)				2.54
16:36:00	59760	175	176	6D	45	not listed	V+(10-14,28-33)	3 (SLS)	1.61E-12	2.76E+12	0.001462	1.48
16:45:00	60300	176	179	7D	45	103,104	V+(10-14,28-33)	3 (SLS)	1.95E-10	1.53E+13	0.014532	1.42
16:55:00	60900	183		8D	45	105,106	V+(10-14,28-33)	3 (SLS)	8.65E-11	1.67E+13	0.006737	3.33
17:08:00	61660	186		9D	45	108,109	V+(10-14,28-33)	3 (SLS)	3.66E-10	6.41E+11	0.002132	4.29
17:19:00	62340	193		10D	45	111,112,113	V+(10-14,28-33)	3 (SLS)				

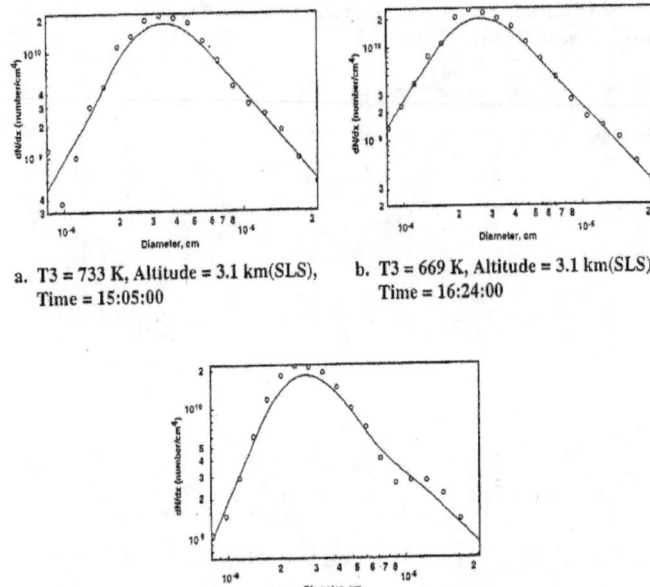

a. T3 = 733 K, Altitude = 3.1 km(SLS),
   Time = 15:05:00

b. T3 = 669 K, Altitude = 3.1 km(SLS),
   Time = 16:24:00

c. T3 = 651 K, Altitude = 3.1 km(SLS),
   Time = 16:36:00

Figure 11. Non-Volatile Aerosol Size Distribution [Howard, 1996]

the Environmental Protection Agency Emission Data Base [EPA, 1985]. The difficult
task of making approximations must somehow be accomplished with very limited
measurement data of the mass emissions. A review of the methods indicates that four
broad categories of estimate methodologies have been used. These are:

- *Simple Factor* multiplied by the number of LTOs.
- The rate of fuel flow multiplied by a *Compound Factor* that includes such
  variables as the ratio of smoke number (aircraft SN of concern compared to an
  aircraft SN with a known mass emission rate), mass measurements (when
  available), thrust, operating pressures and/or temperatures, and other engine
  parameters.
- *Grab Samples and/or Nearby Deposition* to estimate specific emission rates for
  aircraft types or facilities.
- Use of actual *Measured Mass* test results (i.e., USEPA Method 5).

28

3.1 Simple Factor

The first, and most simple approximation is by the use of a multiplying factor applied to the total number of aircraft operations at an airport on any average day. For example, this method has been used to estimate PM emissions at large North American airports. At this airport, the following LTO[8] approximation was used for particulate mass emissions:

- For all General Piston aircraft: 0.01 kg/LTO.
- For all other aircraft, 0.54 kg/LTO.

Exact details of how these values were derived are unavailable. The single factor approximation is very fast and easy to use by avoiding more comprehensive calculations. However, the simple factor method also ignores aircraft fleet mix, changes in aircraft mode times, and other important site-specific variables.

3.2 Compound Factor

In this approximation, a key parameter, usually fuel flow, is allowed to vary and is multiplied by a factor involving many other variables that are held constant [Wayson, 1989]. For example, smoke numbers are available for most engines in the ICAO database and could be directly used. The ICAO database presents smoke numbers for each throttle setting representing the four modes in an LTO cycle. The authors wish to point out that it could be assumed that each of these smoke numbers is a function of mass emitted from an aircraft during that mode. If so, it could be further assumed that a relationship exists between aircraft smoke number, and fuel flow, if engines are somewhat similar. Therefore:

$$EI_i = (SN_i / SN_{ref})(EI_{ref}) \qquad [2]$$

Where:

EI_i = Corrected Emission Index in terms of fuel flow for any aircraft type, i

SN_i = smoke number from ICAO data base for specific aircraft

SN_{ref} = smoke number from a reference aircraft in the ICAO data base

EI_{ref} = known emission Index for reference aircraft

Then:

$$M_{total} = \Sigma_i \Sigma_j (EI_i)(N_i)(F_{ij}) \qquad [3]$$

---

[8] An LTO is a landing/takeoff cycle. It consists of four discrete aircraft operating modes; idle/taxi, takeoff, climbout, and approach. The ICAO has defined standard cycle times for each mode. The four power settings in the ICAO database approximate throttle settings during the four defined aircraft power modes.

Where:

        M_{total} = Total Mass emitted for all aircraft operations
        N_i = Number of Aircraft of type i; and,
        F_{ij} = Fuel Flow for Aircraft type i and mode j.
           (Modes: Idle/Taxi, Takeoff, Climb-out, Approach)

The SN_{ref} from a reference aircraft is that included in the ICAO data base. This reference aircraft type, due to very limited data, must be characteristic of a more general type of engine when approximations are made for all aircraft operations. Most often, this reference mass emission index that has been used is one of the four listed values in the EPA AP-42 database [EPA, 1985]. The four values have been used to represent engines that most resembled the aircraft engine under evaluation. It should be noted that this emission index must be multiplied by the fuel flow, allowing normalization across aircraft types. This permits the fleet mix, changes to the fleet mix, and modal emissions to be considered. In an effort to achieve greater accuracies, other ratios or factors have also been applied for such important considerations as thrust, engine operating temperatures and pressures.

The major weakness of this method is the uncertain correlation between mass and smoke number. The further assumption that smoke numbers can be compared, aircraft-to-aircraft, is even weaker. Consequently, a very low degree of certainty can be associated with this approach.

During a recent environmental analysis performed for Heathrow Airport, an interesting compound factor approach was used [Underwood, 1996]. The approach started with the ICAO Smoke Number database. However, even this database is lacking for certain aircraft types, so, a more universal approach was defined. Using the smoke number for three commonly used engines (RB211-22B, GE CF6-80A, JT9D-7), and the PW at the airport, a representative value for each aircraft mode (take-off, climb-out, approach, and idle) was determined. This representative value was chosen to be conservative so that over-prediction would occur rather than under-prediction. Table 8 shows the smoke number and the representative values selected. All PM was assumed to fall into the PM_{10} category. Next, using the work by Champagne [Champagne, 1971], the representative smoke numbers were correlated to mass emission rates. This provided a concentration of particulate mass in the exhaust stream. But since the emission indexes are desired as a function of fuel flow, the volume of exhaust at Standard Temperature and Pressure (STP) per kilogram of fuel was calculated based on the air to fuel ratio at each thrust (aircraft mode) setting. Again, to be conservative, the upper end of reported air to fuel ratios were used for each aircraft thrust setting (mode). To complete the calculations, jet fuel was assumed to be represented by the chemical formula, $C_{12}H_{22}$, and the stoichiometric relationship was used:

$$C_{12}H_{22} + 17.5\ O_2 \rightarrow 12\ CO_2 + 11\ H_2O \qquad [4]$$

In this case one kilogram of fuel consumes 3.37 kilograms of oxygen, yielding 3.18 kilograms of carbon dioxide and 1.19 kilograms of water. Assuming the density of air to be 1.977 kilograms-per-cubic meter, the volume of exhaust gas in cubic meters can be defined with the following derived equation:

$$\text{Volume of Exhaust} = F/1.293 + 0.72 \qquad [5]$$

Where: F = air-to-fuel ratio

The product of the volume of the exhaust and the representative mass concentration in the exhaust from the Champagne analysis yields the emission factor in the form of kilograms of PM per to kilogram of fuel burned. Table 9 shows the inputs and derived values using this method. Comparison to the AP-42 value for the JT8D shows the derived emission factor for takeoff was within a factor of two. Based on the results of the comparison to the AP-42 levels, the derived factors were then used for all large commercial aircraft at the airport.

Table 8. Representative Values of Smoke Number [Underwood, 1996]

Engine	Smoke Number			
	Take-Off	Climb-Out	Approach	Idle
RB211-22B	14.7	12.9	8.0	3.1
GE CF6-80A	12.0	10.0	2.0	2.0
PW JT9D-7	11.9	9.0	2.5	0.7
Representative Value	20.0	15.0	10.0	5.0

Table 9. Representative Values for Each Mode (Thrust Setting) [Underwood, 1996]

	Take-Off	Climb-Out	Approach	Idle
SN	20.0	15.0	10.0	5.0
Particulate concentrations (mg/m^3)	5.0	4.0	2.5	1.5
Air:fuel ratio	50	60	100	120
Exhaust volume (m^3 at STP) per kg fuel	39	47	78	94
PM_{10} emission factor (10^{-3} kg/kg fuel)	0.20	0.20	0.20	0.14

3.3 Grab Samples and/or Nearby Deposition

In an effort to improve estimate accuracy, some limited sampling (grab sampling) and review of nearby deposition has been performed by various airport agencies and consulting firms. However, the measurement of PM near airports has been, for the most part, inconclusive. A study was performed at Logan International Airport using "advanced chemical fingerprinting" [Ernst, 1997]. This fingerprinting consisted of analysis by gas chromatograph and mass spectrometry to evaluate source-specific chemical combinations. The measurement sites were purposely selected to be very near aircraft operations to help ensure the collected data contained sufficient sample to be meaningful. However, the measurements at these selected sites did not differ significantly from sampling at the background site. Conclusions of the report were that soot deposition was from other urban sources and not from the aircraft. This would tend to indicate that, in urban areas, aircraft operations are only a small portion of the PM deposition.

The authors agree that using the measured data, it could be possible to predict future concentrations of PM by using a form of the rollback model: a ratio of the change in operations would be used to determine future emissions. A major drawback in using this method is the time and site -specific nature of the sample. Flexibility is also limited to allow prediction of emissions by mode or by fleet change. Other nearby sources may also have large impact on estimates, which would require either prediction of these PM emissions to adjust the data for only airport related emissions or only apply to airports that are significantly removed from major nearby PM sources. The use of these adjustments would further limit the confidence placed on estimates.

3.4 Measured Mass Emission Indexes

3.4.1 DLR

DLR has calculated mass emission indexes for carbon black (soot) based on measured data characteristic of jet emissions [Petzold, 1999].
In addition, acknowledging that the ICAO smoke number was the only database that is relatively complete for PM emissions, DLR has developed a methodology to predict the mass emission index using this ICAO reference.

As other researchers had pointed to before, the smoke numbers of different engines vary widely (see Figure 12) [ICAO, 1995]. To overcome this difficulty, DLR used a semi-empirical black carbon correlation method with variable reference values depending on the aircraft type [Dopelheuer, 1997]. The process involves different reference functions for every investigated engine and is based on the SLS black carbon mass concentration versus combustor inlet temperature (T_3). The properties of the fuel, combustor, and injection systems are also considered in an indirect manner. Thermodynamic engine data is used to define the engine operating condition. The process then compares the thrust and smoke number at SLS to the related T_3 and derived carbon black mass emission index. From this, estimates are made of the emission indexes in flight. But of interest in

32

this work is the way the emission index was calculated. Based on the work of Champagne [Champagne, 1971], Hurley [Hurley, 1993], and Whyte [Whyte, 1982], an approximation curve was developed as shown in Figure 13. For mass emission indexes of up to 6 microgram-per-cubic-meter, a combination of the curves presented by Champagne and Hurley used. Above this mass index, the curve presented by Whyte is used. The total black carbon concentration is predicted based on the air and fuel mass flow data and a thermodynamic engine cycle program. The methodology has given promising results. Of note is that a value of 3.2×10^{-17} grams was used to convert between carbon black mass and smoke number.

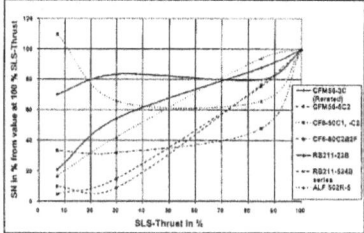

Figure 12. SN Versus SLS-thrust in Percentage Terms for Several Engines [ICAO, 1995]

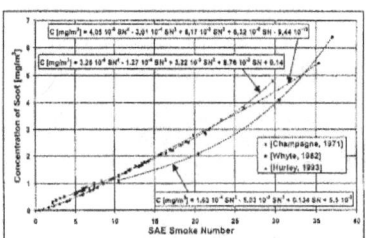

Figure 13. Soot Concentration Versus Smoke Number, Comparison Between Different Authors [Petzold, 1999]

Using the correlation method, Table 10 has been derived. This table, representing nine aircraft and engine combinations, presents data that could be used during airport analysis. Of note is that the same engine does not always have the same emission index for all airframes. This implies that emission indexes for each engine / airframe pair is needed.

33

Table 10. Calculated Black Carbon Mass and Particle Emission Indices for Common Airframes – Engine Combinations for Whole Flight Missions and for 100% Thrust at Sea-Level-Static Conditions [Petzold, 1996]

Airframe	Engine	Engine Certification Year	Engine $EI(BC)$ at 100% SLS Thrust [*]	Flight Mission $EI(BC)$ [*]	Flight Mission $EI(N)$ [+]
B707	JT3D-3B	1962	0.829	0.294	91.2
L1011	RB211-22B	1971	0.089	0.038	11.8
L1011-250	RB211-524B	1975	0.1	0.027	8.8
A300	CF6-50C2	1977	0.02	0.012	3.7
DC 10-30	CF6-50C2	1977	0.02	0.01	3.1
B747-300	CF6-50E2	1977	0.02	0.009	2.8
B747-400	CF6-80C2B1F	1982	0.037	0.016	5.0
B737	CFM56-3B2	1983	0.025	0.013	4.0
MD11	PW 4460	1986	0.043	0.026	8.1
	CFM56-5C2	1990	0.052		

SLS, sea-level-static.
[*] Units are gram of black carbon per kilogram of consumed fuel.
[+] Units are 10^{14} particles per kilogram of consumed fuel.

3.4.2 UMR

UMR has published suggested emission indexes determined during its work. For the work done in conjunction with the AEDC [Howard, 1996], particle indexes were determined. It was found that if particles are sampled at or near the exit plane of the exhaust nozzle, the number of particles per cubic centimeter of exhaust emitted can be determined. Knowledge of the fuel/air ratio relates the mass of fuel use to the particle number count since the volume of fuel is known. The mass indexes were determined during this testing assuming spherical particles and an average density of 2-grams/cubic centimeter. Table 11 shows the pertinent data from the report in regards to emission indexes for PM.

However, as with other databases, the UMR database only includes a sampling of the engines in use. UMR has suggested a more flexible methodology based on correlation of measured particle emission concentrations and their size distributions with the ICAO database smoke numbers [Whitefield, 2001]. Conducted in support of environmental analysis for a major airport, the work by Whitefield and his associates is based on the fact that twenty-nine types of aircraft, with only five major changes in engines, represent 74 percent of total miles flown according to previous global emission inventory data [Gardener, 1998]. Table 12 lists these engine/airframe combinations. The engines include the Rolls Royce RB211, Pratt and Whitney JT8D and JT9D, General Electric CFM56-5C2. The PM emissions were measured using the UMRMASS and results were then correlated with the ICAO database. To perform this correlation for all engines from

34

Table 11. Size Distribution Index and Associated EI's Listed as a Function of Test Parameters [Howard, 1996]

Time (hr:m:s)	Time, sec	Size Distribution Run Numbers Total Aerosol	Non-Volatile	Tank Label	Fill Time, sec	Test Facility Steady-State Data Point	Sample Source (Valve Settings)	Altitude, km	Volume Fraction	EI (Number)	EI (Mass)	%CO₂
7:57:00	28620	1	2	1B(4B)	30	not listed	V+(10-14,28-33)	7.6				3.18
8:22:34	30154	9	10	5A	40	28	V+(10-14,28-33)	7.6	4.33E-11	4.04E+12	0.001494	2.72
8:39:50	31190	12	13	6A	40	30	V+(10-14,28-33)	7.6				
08:44.3	31470			2B	40	32	V+(10-14,28-33)	7.6				
8:59:30	32370			7A	40	not listed	V+(10-14,28-33)	9.1				
9:05:30	32730	21		8A	30	not listed	V+(28-33,10-14)	9.1	3.37E-10	5.93E+12	0.001736	3.10
9:21:24	33684	26	27	9A	40	37,38,40-49	single(V13)	9.1				
9:33:00	34380	29		10A	30	37,38,40-49	V+(28-33,10-14)	9.1	6.98E-09	7.07E+12	0.001703	3.39
9:34:00	34440			11A		37,38,40-49	V+(28-33,10-14)	9.1	6.71E-11	4.59E+12	0.003620	3.42
10:27:20	37640	39		1C	30	37,38,40-49	full across	9.1	2.15E-11	4.59E+12	0.001230	3.42
10:29:00	37740	51	57	2C	30	37,38,40-49	full across	9.1	3.07E-11	9.56E+12	0.003433	3.42
10:42:08	38528	56		3B	30	37,38,40-49	H+(1-6,19-24)	9.1	2.62E-09	8.13E+12	0.028174	3.58
10:47:00	38820	59		4B	30	37,38,40-49	single(V22)	9.1	3.30E-10	6.30E+12	0.002751	3.27
10:54:00	39240	64		5B	30	37,38,40-49	single(V4)	9.1	4.35E-11	9.82E+12	0.005569	3.28
11:00:00	39600	69		6B	30	37,38,40-49	H+(1-6)	9.1	1.15E-10	9.49E+12	0.014340	3.44
11:02:00	39720	76	77	7B	30	37,38,40-49	H+(1-6)	9.1	3.57E-11	7.14E+12	0.003323	3.73
11:16:00	40680	79		8B	30	51,52	V-(28-33)	9.1	4.11E-11	6.84E+09	0.000004	4.11
11:29:00	41340	82		9B	30	54,55	V-(28-33)	9.1	4.15E-11	2.96E+12	0.003198	2.77
11:49:00	42540	87		9B	30	57,58	V-(28-33)	9.1	1.63E-11	6.81E+12	0.002773	3.40
12:07:00	43620	92	93	10-B	30	60-64	V-(28-33)	12.2	6.14E-11	1.44E+13	0.019529	2.31
12:28:00	44880	95		11-B	40	60-64	V13,V31	12.2	4.14E-11	1.01E+13	0.004191	3.70
12:52:00	46320	105	106	1-D	40	66,67,68	V+(10-14,28-33)	12.2	5.49E-11	1.07E+13	0.006893	3.70
12:56:00	46560	108		2-C	50	66,67,68	V+(10-14,28-33)	12.2	2.09E-10	7.46E+12	0.018420	3.70
13:00:00	46800	113		3-C	50	66,67,68	V+(10-14,28-33)	12.2	3.44E-11	8.64E+12	0.015643	2.98
13:20:00	48000	118	119	4-C	50	71,72	V+(10-14,26,33)	12.2	1.43E-11	4.73E+12	0.006778	2.98
13:24:00	48240	125		5-C	50	71,72	V+(10-14,28-33)	12.2				2.14
13:30:00	48600	130		6-C	50	74,75	V+(10-14,28-33)	12.2				
13:45:00	49500	135		7-C	90	not listed	V+(10-14,28-33)	12.2				
14:00:00	50400	138		8-C	60	79	V+(10-14,28-33)	15.2	8.41E-13	1.52E+12	0.000910	3.18
14:08:30	50910	143		9-C	60	not listed	V+(10-14,28-33)	15.2				
14:18:00	51480	151		10-C	60	83,84	V+(10-14,28-33)	15.2	4.34E-11	3.79E+13	0.019356	3.98
14:57:00	53820	156	157	11-C	30	86-94	single(V13)	3(SLS)	1.32E-10	1.55E+13	0.022249	2.43
15:05:00	54300	162	163	12-C	30	86-94	single(V26)	3(SLS)	6.42E-12	1.35E+13	0.000856	3.09
15:53:00	57180	169		1-E	30	86-94	V+(10-14,28-33)	3(SLS)				3.13
15:58:00	57480	172		2E	45	86-94	single(V28)	3(SLS)				3.27
16:07:00	58020	175		4D	45	86-94	single(V22)	3(SLS)				3.37
16:24:00	59040	178	178	5D	45	96,98	single(V11)	3(SLS)	7.03E-11	3.11E+13	0.022435	2.54
16:36:00	59760	179	179	8D	45	not listed	V+(10-14,28-33)	3(SLS)	1.61E-12	2.76E+12	0.001482	1.48
16:45:00	60300	183		7D	45	103,104	V+(10-14,28-33)	3(SLS)	1.95E-10	1.53E+13	0.014532	1.42
16:55:00	60900	188		8O	45	105,106	V+(10-14,28-33)	3(SLS)	8.65E-11	1.67E+13	0.008737	3.33
17:08:00	61650	193		9O	45	108,109	V+(10-14,28-33)	3(SLS)	3.66E-10	6.41E+11	0.002132	4.29
17:19:00	62340	196		10D	45	111,112,113	V+(10-14,28-33)	3(SL5)				
				4E								

the limited measured database several assumptions had to be made:

- Soot particles were assumed to be non-volatile and spherical;
- All particles emitted have a diameter of less than 2.5 micrometers;
- The ratio of non-volatile to total particle mass for the indexes is the same at cruise and sea level;
- Mass-based PM indices are positively correlated with fuel flow;
- The fuel sulfur dependency measured from the RB211 could be applied to all engines; and,
- Correlations can be used for intermediate thrust (throttle) settings between the four included in the ICAO database LTO approach.

The opinion of the authors is that the positive correlation with fuel flow of the mass based PM emissions would seem to be substantiated by UMR when measured data are compared with normalized smoke numbers. A linear slope was shown for this correlation. The normalized smoke number is the smoke number for a given engine divided by the maximum smoke number for that engine as reported in the ICAO database. The slopes determined for this analysis ranged from 0.33 to 0.85. The analysis concluded, "These results show that for the most widely used engines, for which smoke number as a linear function of fuel flow with a positive slope is a reasonable model." The weighted average slope versus fuel flow was found to be 0.613 ± 0.054 s / kilogram.

UMR reasoned that the slope of the normalized smoke number, versus fuel flow, should be the same as the slope for the mass emitted when compared to fuel flow. This is very similar to the analysis method presented in the section on Compound Factor. It was pointed out by UMR that changes to engine components, such as injection nozzles, could have a large impact and invalidate the correlation assumption. A point in case given was that the CF6-50E2 engine with changes in the fuel emission nozzles exhibits completely different characteristics versus the CF6-50C engine. Emissions may even be independent of fuel flow. The degree of uncertainty is high when applied in a general case to all engines and a general error of up to 30 percent was reported by the authors.

The exact methodology was then given as:

$$fei = EI_{UMR} [1+(0.613 \text{ s/kg})(ff - ff_{UMR})] \qquad [6]$$

Where:
fei = linearily fitted PM emission index for engine i;
EI_{UMR} = emission index measured by UMR;
ff = fuel flow [kg/s]; and,
ff_{UMR} = fuel flow for measured EI at UMR.

Table 12. Usage and Relationship Applicability Data for 29 of the Most Widely Used Engines in the Commercial Fleet, Based on Distances Flown [Whitefield, 2001]

aircraft	distance	CFM56B1	CFM56B2	CFM56C2	PWJT8D7	PWJT8D9
727	2.43E+06				7.9	25.3
737 100/200	1.78E+06				7.2	31.3
737 300/400	1.53E+06	45.4	36.9	17.5		
747 100/300	2.18E+06					
767	1.73E+06					
DC10	1.53E+06					
DC9	1.21E+06				44.3	33.2
MD80	1.40E+06					
total	1.73E+07					
de		6.937E+05	5.638E+05	2.674E+05	8.553E+05	1.574E+06
Re		0.040	0.033	0.015	0.050	0.091
slope		0.39292	0.79406	0.57655	X	X

aircraft	distance	PWJT8D17C	PWJT8D217	PWJT8D219	GECF650C	GECF650C1
727	2.43E+06	7.6				
737 100/200	1.78E+06					
737 300/400	1.53E+06					
747 100/300	2.18E+06					17.6
767	1.73E+06					
DC10	1.53E+06				11	34.1
DC9	1.21E+06					
MD80	1.40E+06		56.9	39.7		
total	1.73E+07					
de		1.846E+05	7.982E+05	5.569E+05	1.679E+05	9.034E+05
Re		0.011	0.046	0.032	0.010	0.052
slope		X	X	X	Y	0.87464

aircraft	distance	GECF680C2B6	GECF680C2D1F	PW4060	PW4460	PWJT9D7A
727	2.43E+06					
737 100/200	1.78E+06					
737 300/400	1.53E+06					
747 100/300	2.18E+06					31.1
767	1.73E+06	11.4		16		
DC10	1.53E+06		7.5		5.4	5.1
DC9	1.21E+06					
MD80	1.40E+06					
total	1.73E+07					
de		1.968E+05	1.145E+05	2.762E+05	8.241E+04	7.546E+05
Re		0.011	0.007	0.016	0.005	0.044
slope		0.52356	0.51839	X	X	0.49459

Next, the actual mass-based emission factor was calculated by:

$$fei\_m = (\pi/6) \: \rho \: (Xbarv)^3 \: (fei) \qquad\qquad [7]$$

Where:
fei_m = mass based EI;
ρ = 1.0 gram/cubic centimeter
Xbarv = mean volumetric diameter for the measured particle size distribution.

Table 13 shows the results from this method. Note that final EIs include an enhancement correction factor for the fuel sulfur content based on the testing of the RB211-535C engine.

The similarities between this method and the Compound Factor method are substantial and occur multiple times. Total mass is then predicted the same way as the Compound Factor.

3.4.3 United States Navy

The U.S. Navy has performed testing to characterize aircraft particulate emissions [AESO, 1990] using the EPA Method 5 approach for engines at idle, power settings higher than idle (e.g., 30 percent thrust or 85 percent RPM), and the military setting of intermediate rated power (IRP). Information collected in addition to PM included engine type, number of tests performed to obtain average values, fuel flow rate, type of fuel used with heat of combustion of the fuel, stack gas temperature, stack gas flow rates, and carbon dioxide concentration in the stack gas. It was noted that adjustment of the sample to a carbon dioxide content of 1 percent allows comparison of the adjusted PM concentrations to the relevant discharge limit, without bias from the test cell design and the impact of excess air.

The report lists particulate emissions from the front half of the sampling system and the total collected. The front half of the system is considered to be the particulate mass captured in the heated sampling probe and on the filter during Method 5 testing. Front half emissions account for material that would exist in PM form at about 248 degrees Fahrenheit. The back half accounts for the remaining PM collected in the impingers at room temperature (about 68 degrees Fahrenheit). The condensable matter, captured in the back half of sampling, can contribute significantly to the total mass of PM emissions.

Table 13. Estimates of Number- and Mass-Based EI's for the LTO Cycle of Four Popular Engines [Whitefield, 2001]

Engine: RB211-22B Xbarv=47.8nm				fei_m	
Cond	ff	fei	NV	Tot_HS	Tot_LS
	(kg/s)	(#/kg_f)	(g/kg_f)	(g/kg_f)	(g/kg_f)
T/O	1.9	1.3E+15	0.074	1.1	0.19
C/O	1.5	1.2E+15	0.066	1.0	0.17
App	0.55	7.3E+14	0.042	0.61	0.11
Idle	0.28	6.2E+14	0.035	0.52	0.091

Engine: JT9D-7J Xbarv=106nm				fei_m	
Cond	ff	fei	NV	Tot_HS	Tot_LS
	(kg/s)	(#/kg_f)	(g/kg_f)	(g/kg_f)	(g/kg_f)
T/O	2.3	8.4E+14	0.52	7.7	1.3
C/O	1.9	7.2E+14	0.45	6.6	1.2
App	0.68	3.9E+14	0.24	3.5	0.62
Idle	0.24	2.6E+14	0.17	2.4	0.43

Engine: CFM56-5C2 Xbarv=107nm				fei_m	
Cond	ff	fei	NV	Tot_HS	Tot_LS
	(kg/s)	(#/kg_f)	(g/kg_f)	(g/kg_f)	(g/kg_f)
T/O	1.3	1.2E+15	0.80	12	2.1
C/O	1.1	1.0E+15	0.65	10	1.7
App	0.36	3.1E+14	0.20	2.9	0.52
Idle	0.12	7.7E+13	0.050	0.73	0.13

Engine: JT8D Xbarv=149nm				fei_m	
Cond	ff	fei	NV	Tot_HS	Tot_LS
	(kg/s)	(#/kg_f)	(g/kg_f)	(g/kg_f)	(g/kg_f)
T/O	1.3	2.5E+15	4.2	62	11
C/O	1.0	2.2E+15	3.8	56	10
App	0.35	1.5E+15	2.7	39	6.9
Idle	0.15	1.3E+15	2.3	34	6.0

Tables 14 through 26, from this data collection effort, show the measured results in:

- grains[9] per cubic foot (PM captured from sampling at the exhaust plane of the test cell stack, uncorrected for temperature, pressure, carbon dioxide concentration, or moisture content);
- grains per standard cubic foot (adjusted to a moisture free condition at standard temperature and pressure[10]);
- grains per standard cubic foot adjusted for a carbon dioxide concentration of one percent;
- grains per standard cubic foot adjusted for a carbon dioxide concentration of twelve percent (a common practice in fossil fuel power plant PM measurements); and,
- emission rates in pounds/hour, pounds/million BTU, and pounds/thousand pounds of fuel are applicable for actual engine operating conditions.

Engines tested for PM included the J79-GE-8C, J79-GE-8D, J79-GE-10B, J52-P-6B, TF30-P-414, F404-GE-400, TF34-GE-400A, T58-GE-5/8F, and T64-GE-6B/415. Unfortunately, most of these engines are not used in commercial aircraft and many have very different operating conditions from those used in such aircraft.

As previously mentioned, emission rates were also determined. Figure 14 and 15 show the results for emission indexes versus fuel flow rate. Figure 16 shows the results of total PM emissions versus fuel flow rate.

---

[9] A grain is 1/7000th of a pound and is commonly used in particulate investigations.
[10] Standard temperature is 68 degrees Fahrenheit and standard pressure is 29.92 inches of mercury.

Table 14. Particulate Mass Emissions From the J79-GE-8C/8D Engine; Summary Files 1-7 [AESO, 1990]

	ENGINE POWER SETTING	
	Idle	30% Thrust
Number of tests	7	10
Fuel flow rate (lbs/hr)	1189	2893
Fuel type: JP-5		
Heat of combustion (Btu/lb): 18500		
STACK GAS DATA		
Temperature (°F)	230.3	225.7
Actual flow rate (cfm)	205250	445051
Corrected flow rate (cfm, dry, corrected to to standard conditions)	154080	336154
Carbon dioxide (volume percent, dry basis)	0.54	0.51
PARTICULATE EMISSIONS DATA		
Partial emissions (front half)		
grain/cu ft at stack conditions	0.008	0.009
grain/std cu ft, dry	0.010	0.012
grain/std cu ft, dry, corrected to 1% CO_2	0.019	0.024
grain/std cu ft, dry, corrected to 12% CO_2	0.230	0.289
pound/hour	13.69	34.45
pound/million Btu	0.62	0.64
pound/thousand pound fuel	11.55	11.93
Total emissions		
grain/cu ft at stack conditions	0.013	0.012
grain/std cu ft, dry	0.017	0.015
grain/std cu ft, dry, corrected to 1% CO_2	0.032	0.031
grain/std cu ft, dry, corrected to 12% CO_2	0.385	0.369
pound/hour	22.69	44.41
pound/million Btu	1.03	0.83
pound/thousand pound fuel	19.12	15.34

Note: All entries are averages, derived directly from the individual test
      files in Appendix A, ref. 2. The data were obtained at Test Cell 19,
      Building 397, Naval Air Rework Facility, North Island.

Table 15. Particulate Mass Emissions From the J79-GE-8C/8D Engine; Summary Files 18-27 [AESO, 1990]

	ENGINE POWER SETTING
	Military
Number of tests	10
Fuel flow rate (lbs/hr)	9259
Fuel type: JP-5	
Heat of combustion (Btu/lb): 18500	
STACK GAS DATA	
Temperature (°F)	395
Actual flow rate (cfm)	863488
Corrected flow rate (cfm, dry, corrected to to standard conditions)	522431
Carbon dioxide (volume percent, dry basis)	1.05
PARTICULATE EMISSIONS DATA	
Partial emissions (front half)	
grain/cu ft at stack conditions	0.009
grain/std cu ft, dry	0.015
grain/std cu ft, dry, corrected to 1% CO_2	0.014
grain/std cu ft, dry, corrected to 12% CO_2	0.171
pound/hour	67.27
pound/million Btu	0.39
pound/thousand pound fuel	7.29
Total emissions	
grain/cu ft at stack conditions	0.013
grain/std cu ft, dry	0.022
grain/std cu ft, dry, corrected to 1% CO_2	0.021
grain/std cu ft, dry, corrected to 12% CO_2	0.251
pound/hour	98.51
pound/million Btu	0.58
pound/thousand pound fuel	10.66

Note: All entries are averages, derived directly from the individual test
      files in Appendix A, ref. 2. The data were obtained at Test Cell 19,
      Building 397, Naval Air Rework Facility, North Island.

Table 16. Particulate Mass Emissions From the J79-GE-8D Engine; Summary Files 28-33 [AESO, 1990]

	ENGINE POWER SETTING
	Military
Number of tests Fuel flow rate (lbs/hr) Fuel type: JP-5 Heat of combustion (Btu/lb): 18500	6 9366
STACK GAS DATA	
Temperature (°F) Actual flow rate (cfm) Corrected flow rate (cfm, dry, corrected to to standard conditions) Carbon dioxide (volume percent, dry basis)	404.2 789586 463573 1.07
PARTICULATE EMISSIONS DATA	
Partial emissions (front half)	

grain/cu ft at stack conditions
  grain/std cu ft, dry
  grain/std cu ft, dry, corrected to 1% CO_2
  grain/std cu ft, dry, corrected to 12% CO_2
  pound/hour
  pound/million Btu
  pound/thousand pound fuel
	0.010
	0.017
	0.016
	0.187
	65.58
	0.38
	7.03

Total emissions	

grain/cu ft at stack conditions
  grain/std cu ft, dry
  grain/std cu ft, dry, corrected to 1% CO_2
  grain/std cu ft, dry, corrected to 12% CO_2
  pound/hour
  pound/million Btu
  pound/thousand pound fuel
	0.015
	0.025
	0.024
	0.284
	99.96
	0.58
	10.68

Note: All entries are averages, derived directly from the individual test
      files in Appendix A, ref. 2. The data were obtained at Test Cell A,
      Building 545, NAS Miramar.

Table 17. Particulate Mass Emissions From the J79-GE-8C/8D Engine; Summary Files 1-33 [AESO, 1990]

TABLE 4.   PARTICULATE MASS EMISSIONS FROM THE J79–GE–8C/8D ENGINE
(Summary of Files 1 through 33)

	ENGINE POWER SETTING
	Military
Number of tests	16
Fuel flow rate (lbs/hr)	9299
Fuel type: JP-5	
Heat of combustion (Btu/lb): 18500	
STACK GAS DATA	
Temperature (°F)	N/A
Actual flow rate (cfm)	N/A
Corrected flow rate (cfm, dry, corrected to to standard conditions)	N/A
Carbon dioxide (volume percent, dry basis)	N/A
PARTICULATE EMISSIONS DATA	
Partial emissions (front half)	
grain/cu ft at stack conditions	0.009
grain/std cu ft, dry	0.016
grain/std cu ft, dry, corrected to 1% CO_2	0.015
grain/std cu ft, dry, corrected to 12% CO_2	0.177
pound/hour	66.64
pound/million Btu	0.39
pound/thousand pound fuel	7.19
Total emissions	
grain/cu ft at stack conditions	0.014
grain/std cu ft, dry	0.023
grain/std cu ft, dry, corrected to 1% CO_2	0.022
grain/std cu ft, dry, corrected to 12% CO_2	0.263
pound/hour	99.05
pound/million Btu	0.58
pound/thousand pound fuel	10.67

Note: All entries are averages, derived directly from the individual test
      files in Appendix A, ref. 2. The data were obtained at Test Cell 19,
      Building 397, Naval Air Rework Facility, North Island and at Test
      Cell A, Building 545, NAS Miramar.

Table 18. Particulate Mass Emissions From the J79-GE-8D Engine with Ferrocene
Additive; Summary Files 34-45 [AESO, 1990]

	ENGINE POWER SETTING
	Military
Number of tests	12
Fuel flow rate (lbs/hr)	9243
Fuel type: JP-5	
Heat of combustion (Btu/lb): 18500	
STACK GAS DATA	
Temperature (°F)	395
Actual flow rate (cfm)	835800
Corrected flow rate (cfm, dry, corrected to to standard conditions)	504700
Carbon dioxide (volume percent, dry basis)	1.01
PARTICULATE EMISSIONS DATA	
Partial emissions (front half)	
grain/cu ft at stack conditions	0.008
grain/std cu ft, dry	0.013
grain/std cu ft, dry, corrected to 1% CO_2	0.014
grain/std cu ft, dry, corrected to 12% CO_2	0.170
pound/hour	59.60
pound/million Btu	0.35
pound/thousand pound fuel	6.45
Total emissions	
grain/cu ft at stack conditions	0.015
grain/std cu ft, dry	0.024
grain/std cu ft, dry, corrected to 1% CO_2	0.025
grain/std cu ft, dry, corrected to 12% CO_2	0.304
pound/hour	105.10
pound/million Btu	0.62
pound/thousand pound fuel	11.37

Note: All entries are averages, derived directly from the individual test
files in Appendix A, ref. 2. The data were obtained at Test Cell 19,
Building 397, Naval Air Rework Facility, North Island.

Table 19. Particulate Mass Emissions From the J79-GE-8D Engine with Ferrocene
Additive; Summary Files 46-47 [AESO, 1990]

	ENGINE POWER SETTING
	Military
Number of tests	2
Fuel flow rate (lbs/hr)	9243
Fuel type: JP-5	
Heat of combustion (Btu/lb): 18500	
STACK GAS DATA	
Temperature (°F)	416.4
Actual flow rate (cfm)	824326
Corrected flow rate (cfm, dry, corrected to	479092
to standard conditions)	
Carbon dioxide (volume percent, dry basis)	1.06
PARTICULATE EMISSIONS DATA	
Partial emissions (front half)	
grain/cu ft at stack conditions	0.005
grain/std cu ft, dry	0.008
grain/std cu ft, dry, corrected to 1% CO_2	0.008
grain/std cu ft, dry, corrected to 12% CO_2	0.095
pound/hour	33.92
pound/million Btu	0.20
pound/thousand pound fuel	3.67
Total emissions	
grain/cu ft at stack conditions	0.007
grain/std cu ft, dry	0.012
grain/std cu ft, dry, corrected to 1% CO_2	0.011
grain/std cu ft, dry, corrected to 12% CO_2	0.132
pound/hour	47.10
pound/million Btu	0.28
pound/thousand pound fuel	5.10

Note: All entries are averages, derived directly from the individual test
      files in Appendix A, ref. 2. The data were obtained at Test Cell A,
      Building 545, NAS Miramar

Table 20. Particulate Mass Emissions From the J79-GE-8D Engine with Ferrocene Additive; Summary Files 34-47 [AESO, 1990]

	ENGINE POWER SETTING
	Military
Number of tests	14
Fuel flow rate (lbs/hr)	9243
Fuel type: JP-5	
Heat of combustion (Btu/lb): 18500	
STACK GAS DATA	
Temperature (°F)	N/A
Actual flow rate (cfm)	N/A
Corrected flow rate (cfm, dry, corrected to to standard conditions)	N/A
Carbon dioxide (volume percent, dry basis)	N/A
PARTICULATE EMISSIONS DATA	
Partial emissions (front half)	
grain/cu ft at stack conditions	0.008
grain/std cu ft, dry	0.012
grain/std cu ft, dry, corrected to 1% CO_2	0.013
grain/std cu ft, dry, corrected to 12% CO_2	0.159
pound/hour	55.93
pound/million Btu	0.33
pound/thousand pound fuel	6.05
Total emissions	
grain/cu ft at stack conditions	0.014
grain/std cu ft, dry	0.022
grain/std cu ft, dry, corrected to 1% CO_2	0.023
grain/std cu ft, dry, corrected to 12% CO_2	0.279
pound/hour	96.81
pound/million Btu	0.56
pound/thousand pound fuel	10.47

Note: All entries are averages, derived directly from the individual test files in Appendix A, ref. 2. The data were obtained at Test Cell 19, Building 397, Naval Air Rework Facility, North Island and at Test Cell A, Building 545, NAS Miramar.

Table 20. Particulate Mass Emissions From the J79-GE-8C/8D Engine; Summary of Table 2, 3, 5 and 6 [AESO, 1990]

	ENGINE POWER SETTING
	Military
Number of tests	30
Fuel flow rate (lbs/hr)	9273
Fuel type: JP-5	
Heat of combustion (Btu/lb): 18500	
STACK GAS DATA	
Temperature (°F)	N/A
Actual flow rate (cfm)	N/A
Corrected flow rate (cfm, dry, corrected to to standard conditions)	N/A
Carbon dioxide (volume percent, dry basis)	N/A
PARTICULATE EMISSIONS DATA	
Partial emissions (front half)	
grain/cu ft at stack conditions	0.009
grain/std cu ft, dry	0.014
grain/std cu ft, dry, corrected to 1% CO_2	0.014
grain/std cu ft, dry, corrected to 12% CO_2	0.169
pound/hour	61.64
pound/million Btu	0.36
pound/thousand pound fuel	6.66
Total emissions	
grain/cu ft at stack conditions	0.014
grain/std cu ft, dry	0.023
grain/std cu ft, dry, corrected to 1% CO_2	0.023
grain/std cu ft, dry, corrected to 12% CO_2	0.271
pound/hour	98.01
pound/million Btu	0.57
pound/thousand pound fuel	10.58

Note: All entries are averages, derived directly from the individual test files in Appendix A, ref. 2. The data were obtained at Test Cell 19, Building 397, Naval Air Rework Facility, North Island and at Test Cell A, Building 545, NAS Miramar.

Table 21. Particulate Mass Emissions From the J79-GE-10B Engine; Summary of Files 48-65 [AESO, 1990]

	ENGINE POWER SETTING		
	Idle	30% Thrust	Military
Number of tests	6	7	5
Fuel flow rate (lbs/hr)	1249	2954	9886
Fuel type: JP-5			
Heat of combustion (Btu/lb): 18500			
STACK GAS DATA			
Temperature (°F)	223.9	225.7	413.6
Actual flow rate (cfm)	197418	431380	859372
Corrected flow rate (cfm, dry, corrected to to standard conditions)	149861	325466	512129
Carbon dioxide (volume percent, dry basis)	0.55	0.52	1.06
PARTICULATE EMISSIONS DATA			
Partial emissions (front half)			
grain/cu ft at stack conditions	0.006	0.005	0.004
grain/std cu ft, dry	0.008	0.006	0.006
grain/std cu ft, dry, corrected to 1% CO_2	0.014	0.012	0.006
grain/std cu ft, dry, corrected to 12% CO_2	0.170	0.144	0.070
pound/hour	9.82	17.33	26.56
pound/million Btu	0.42	0.32	0.15
pound/thousand pound fuel	7.87	5.84	2.69
Total emissions			
grain/cu ft at stack conditions	0.012	0.008	0.006
grain/std cu ft, dry	0.015	0.010	0.010
grain/std cu ft, dry, corrected to 1% CO_2	0.028	0.019	0.010
grain/std cu ft, dry, corrected to 12% CO_2	0.336	0.234	0.116
pound/hour	19.62	28.12	43.81
pound/million Btu	0.85	0.51	0.24
pound/thousand pound fuel	15.73	9.50	4.43

Note: All entries are averages, derived directly from the individual test files in Appendix A, ref. 2.

Table 21. Particulate Mass Emissions From the J52-P-6B Engine; Summary of Files 66-77 [AESO, 1990]

	ENGINE POWER SETTING		
	Idle	30% rpm	Military
Number of tests	5	4	3
Fuel flow rate (lbs/hr)	849	2660	6765
Fuel type: JP-5			
Heat of combustion (Btu/lb): 18500			
STACK GAS DATA			
Temperature (°F)	167.6	213.6	325.4
Actual flow rate (cfm)	172104	416136	709725
Corrected flow rate (cfm, dry, corrected to to standard conditions)	140118	317962	458079
Carbon dioxide (volume percent, dry basis)	0.34	0.46	0.77
PARTICULATE EMISSIONS DATA			
Partial emissions (front half)			
grain/cu ft at stack conditions	0.007	0.007	0.005
grain/std cu ft, dry	0.009	0.009	0.007
grain/std cu ft, dry, corrected to 1% CO_2	0.027	0.019	0.010
grain/std cu ft, dry, corrected to 12% CO_2	0.327	0.227	0.116
pound/hour	11.08	23.43	29.12
pound/million Btu	0.70	0.47	0.23
pound/thousand pound fuel	13.00	8.78	4.30
Total emissions			
grain/cu ft at stack conditions	0.011	0.010	0.009
grain/std cu ft, dry	0.014	0.013	0.013
grain/std cu ft, dry, corrected to 1% CO_2	0.042	0.028	0.017
grain/std cu ft, dry, corrected to 12% CO_2	0.498	0.335	0.207
pound/hour	16.91	34.90	52.13
pound/million Btu	1.08	0.71	0.42
pound/thousand pound fuel	19.94	13.13	7.75

Note: All entries are averages, derived directly from the individual test
      files in Appendix A, ref. 2.

Table 22. Particulate Mass Emissions From the TF30-P-414 Engine; Summary of Files 78-94 [AESO, 1990]

	ENGINE POWER SETTING		
	Idle	85% rpm	Military
Number of tests	6	6	5
Fuel flow rate (lbs/hr)	1040	2885	8009
Fuel type: JP-5			
Heat of combustion (Btu/lb): 18500			
STACK GAS DATA			
Temperature (°F)	159.2	200.0	324.7
Actual flow rate (cfm)	189575	436292	795029
Corrected flow rate (cfm, dry, corrected to to standard conditions)	157299	340143	517203
Carbon dioxide (volume percent, dry basis)	0.35	0.46	0.76
PARTICULATE EMISSIONS DATA			
Partial emissions (front half)			
grain/cu ft at stack conditions	0.003	0.004	0.002
grain/std cu ft, dry	0.004	0.005	0.003
grain/std cu ft, dry, corrected to 1% CO_2	0.010	0.010	0.004
grain/std cu ft, dry, corrected to 12% CO_2	0.123	0.124	0.052
pound/hour	4.78	13.78	14.79
pound/million Btu	0.25	0.26	0.10
pound/thousand pound fuel	4.60	4.76	1.85
Total emissions			
grain/cu ft at stack conditions	0.006	0.006	0.003
grain/std cu ft, dry	0.007	0.008	0.005
grain/std cu ft, dry, corrected to 1% CO_2	0.020	0.017	0.007
grain/std cu ft, dry, corrected to 12% CO_2	0.239	0.208	0.084
pound/hour	9.35	22.85	23.86
pound/million Btu	0.48	0.43	0.16
pound/thousand pound fuel	8.96	7.98	2.98

Note: All entries are averages, derived directly from the individual test
      files in Appendix A, ref. 2.

Table 23. Particulate Mass Emissions From the F404-GE-400 Engine; Summary of Files 95-105 [AESO, 1990]

	ENGINE POWER SETTING		
	Idle	86% rpm	Military
Number of tests	5	2	4
Fuel flow rate (lbs/hr)	835	3557	8579
Fuel type: JP-5			
Heat of combustion (Btu/lb): 18500			
STACK GAS DATA			
Temperature (°F)	173.5	257.8	381.9
Actual flow rate (cfm)	193418	537300	837083
Corrected flow rate (cfm, dry, corrected to to standard conditions)	159351	389954	515991
Carbon dioxide (volume percent, dry basis)	0.45	0.82	0.93
PARTICULATE EMISSIONS DATA			
Partial emissions (front half)			
grain/cu ft at stack conditions	0.004	0.002	0.003
grain/std cu ft, dry	0.004	0.002	0.004
grain/std cu ft, dry, corrected to 1% CO_2	0.010	0.003	0.004
grain/std cu ft, dry, corrected to 12% CO_2	0.121	0.032	0.053
pound/hour	6.10	7.34	18.44
pound/million Btu	0.40	0.11	0.12
pound/thousand pound fuel	7.30	2.06	2.15
Total emissions			
grain/cu ft at stack conditions	0.006	0.005	0.003
grain/std cu ft, dry	0.007	0.006	0.005
grain/std cu ft, dry, corrected to 1% CO_2	0.017	0.008	0.006
grain/std cu ft, dry, corrected to 12% CO_2	0.199	0.094	0.070
pound/hour	10.34	21.72	24.10
pound/million Btu	0.67	0.33	0.16
pound/thousand pound fuel	12.38	6.10	2.81

Note: Fuel flow rates are representative, All entries are averages, directly from the individual files in Appendix A, ref. 2.

Table 24. Particulate Mass Emissions From the TF34-GE-400A Engine; Summary of Files 106-113 [AESO, 1990]

	ENGINE POWER SETTING		
	Idle	75%	94%
Number of tests	5	3	2
Fuel flow rate (lbs/hr)	450	500	2805
Fuel type: JP-5			
Heat of combustion (Btu/lb): 18500			
STACK GAS DATA			
Temperature (°F)	101.2	103	146
Actual flow rate (cfm)	178438	325658	759800
Corrected flow rate (cfm, dry, corrected to to standard conditions)	166756	305055	679100
Carbon dioxide (volume percent, dry basis)	0.16	0.16	0.24
PARTICULATE EMISSIONS DATA			
Partial emissions (front half)			
grain/cu ft at stack conditions	N/A	N/A	N/A
grain/std cu ft, dry	N/A	N/A	N/A
grain/std cu ft, dry, corrected to 1% CO_2	N/A	N/A	N/A
grain/std cu ft, dry, corrected to 12% CO_2	N/A	N/A	N/A
pound/hour	N/A	N/A	N/A
pound/million Btu	N/A	N/A	N/A
pound/thousand pound fuel	N/A	N/A	N/A
Total emissions			
grain/cu ft at stack conditions	0.001	0.001	-
grain/std cu ft, dry	0.001	0.001	0.001
grain/std cu ft, dry, corrected to 1% CO_2	0.007	0.008	0.004
grain/std cu ft, dry, corrected to 12% CO_2	0.082	0.100	0.005
pound/hour	1.47	3.43	5.92
pound/million Btu	0.18	0.37	0.11
pound/thousand pound fuel	3.26	6.85	2.11

Note: All entries are averages, derived directly from the individual test
      files in Appendix A, ref. 2. The data were obtained at the AIMD Test
      Cell, Naval Air Station, North Island.

Table 25. Particulate Mass Emissions From the T58-GE-5/8F Engine; Summary of Files 114-115 [AESO, 1990]

	ENGINE POWER SETTING
	Various
Number of tests	2
Fuel flow rate (lbs/hr)	423
Fuel type: JP-5	
Heat of combustion (Btu/lb): 18500	
STACK GAS DATA	
Temperature (°F)	500.8
Actual flow rate (cfm)	N/A
Corrected flow rate (cfm, dry, corrected to to standard conditions)	12023
Carbon dioxide (volume percent, dry basis)	1.50
PARTICULATE EMISSIONS DATA	
Partial emissions (front half)	
grain/cu ft at stack conditions	N/A
grain/std cu ft, dry	0.011
grain/std cu ft, dry, corrected to 1% CO_2	0.007
grain/std cu ft, dry, corrected to 12% CO_2	0.085
pound/hour	1.06
pound/million Btu	0.14
pound/thousand pound fuel	2.54
Total emissions	
grain/cu ft at stack conditions	N/A
grain/std cu ft, dry	0.018
grain/std cu ft, dry, corrected to 1% CO_2	0.012
grain/std cu ft, dry, corrected to 12% CO_2	0.104
pound/hour	1.76
pound/million Btu	0.23
pound/thousand pound fuel	4.20

Note: All entries are averages, derived from the individual test files in Appendix A, ref. 2. The data were obtained at Test Cell 11 at Naval Air Rework Facility, North Island as preliminary source tesing.

Table 26. Particulate Mass Emissions From the T64-GE-6B/415 Engine; Summary of Files 116-117 [AESO, 1990]

	ENGINE POWER SETTING
	Various
Number of tests	2
Fuel flow rate (lbs/hr)	1006
Fuel type: JP-5	
Heat of combustion (Btu/lb): 18500	
STACK GAS DATA	
Temperature (°F)	449.2
Actual flow rate (cfm)	N/A
Corrected flow rate (cfm, dry, corrected to	19865
to standard conditions)	
Carbon dioxide (volume percent, dry basis)	2.20
PARTICULATE EMISSIONS DATA	
Partial emissions (front half)	
grain/cu ft at stack conditions	N/A
grain/std cu ft, dry	0.008
grain/std cu ft, dry, corrected to 1% CO_2	0.004
grain/std cu ft, dry, corrected to 12% CO_2	0.042
pound/hour	1.42
pound/million Btu	0.07
pound/thousand pound fuel	1.33
Total emissions	
grain/cu ft at stack conditions	N/A
grain/std cu ft, dry	0.013
grain/std cu ft, dry, corrected to 1% CO_2	0.006
grain/std cu ft, dry, corrected to 12% CO_2	0.070
pound/hour	2.34
pound/million Btu	0.12
pound/thousand pound fuel	2.21

Note: All entries are averages, derived from the individual test files in
      Appendix A, ref. 2. The data were obtained at Test Cells 9 and 10 at
      Naval Air Rework Facility, North Island as preliminary source tesing.

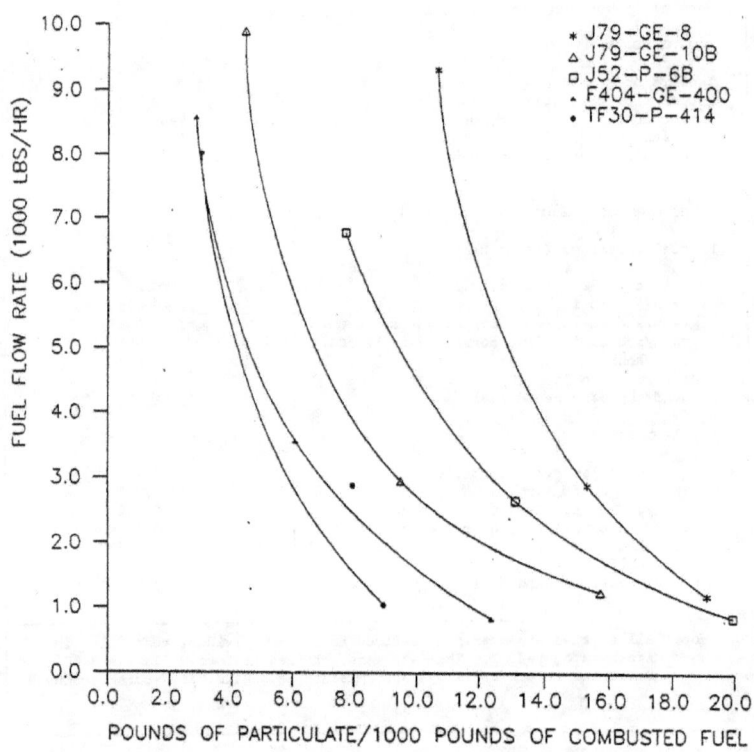

Figure 14. Fuel Flow Rate Vs. Emission Index for Military Gas Turbine Engines [AESO, 1990]

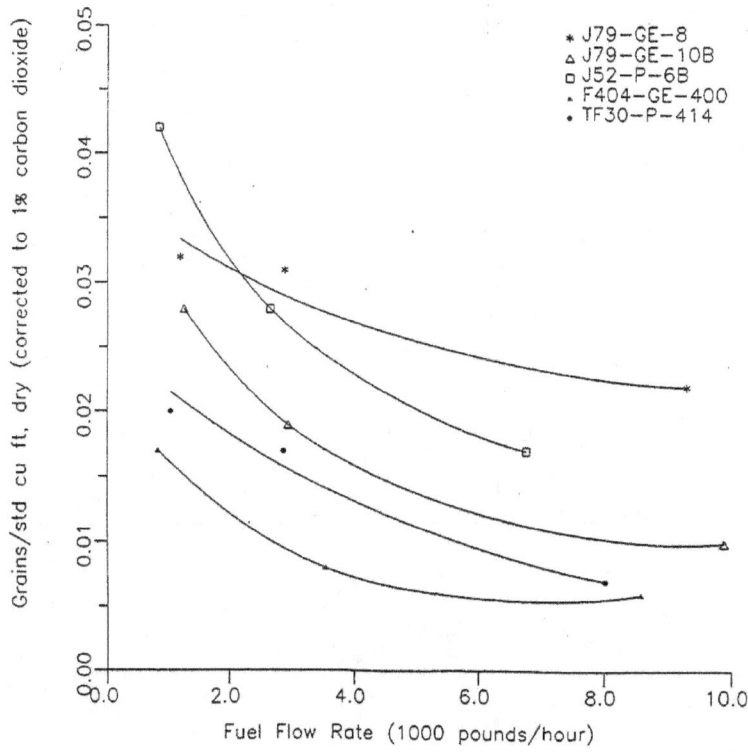

Figure 15. Total Particulate Emissions Vs. Fuel Flow Rate for Military Gas Turbine
Engines [AESO, 1990]

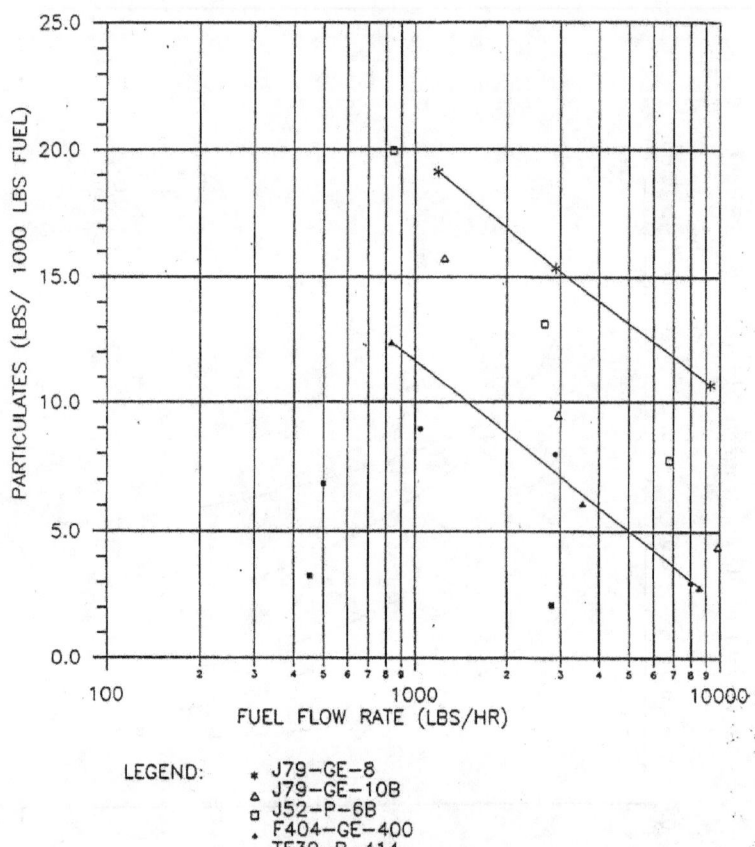

Figure 16. Emission Index Vs. Fuel Flow Rate [AESO, 1990]

3.4.4 United States Air Force

The U.S. Air Force has conducted PM test cell measurements of multiple aircraft engines, two helicopter engines, and two auxiliary power units [EQM, 1998]. A typical test cell configuration used for the aircraft engine testing that was used at Kelly Air Force Base, is shown in Figure 17. Figure 18 shows a typical test cell configuration for auxiliary power units, also at Kelly Air Force Base. Unfortunately, this two-year testing program performed at multiple locations, was done using JP-8 fuel. Unlike JP-4, which is a naptha based

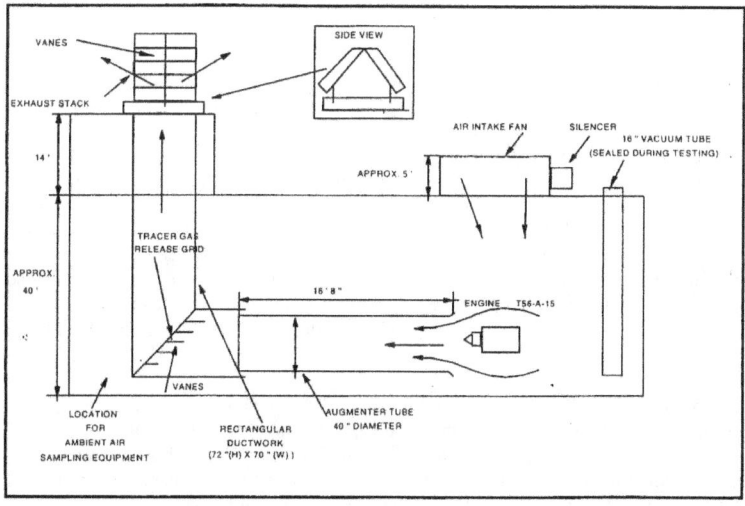

Figure 17. Test Cell 54A Kelly AFB [EQM, 1998]

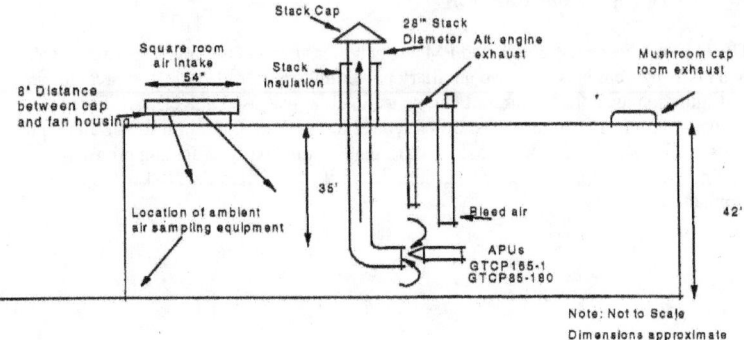

Figure 18. APU Test Facility Kelly AFB [EQM, 1998]

fuel commonly used for commercial aircraft, JP-8 is kerosene based. This leads to differences in the emissions. Table 27 lists the different engines tested, aircraft type, location of tests, and the test dates.

Testing for most criteria pollutants, including PM, was accomplished by Environmental Quality Management, Inc, and Roy F. Weston, Inc. Sampling was not performed for sulfur dioxide. For PM, a modified EPA Method 5 methodology was used. The primary modification to the Method 5 procedure was that the sample was only taken from a single point. This differs considerably from the findings of the Navy, which strongly indicated that full traverse[11] sampling was required for PM. However, the Air Force report stated, "A verification was made through the use of tracer gas that the sample point was representative of the entire exhaust stream." In addition to EPA Method 5, EPA Method 202 was also used. This permitted both filterable and condensable PM to be measured.

Also performed was testing for several metals and compounds including antimony, arsenic, barium, beryllium, cadmium, cobalt, chromium, copper, lead, manganese, mercury, nickel, phosphorus, selenium, silver, thallium and zinc. All were found to be below the minimum detection limit. This seem to indicate to the authors of this review that emissions of metals as PM may be neglected in a first approximation. Total flow rates were not measured directly, but were

---

[11] Traverse sampling is conducted by extracting samples at several predefined locations across the entire exhaust stream to avoid bias at any one location. If complete mixing occurs, such as for gases, full traverse sampling may not be required. As such, if sampling is done at only one location for PM it is assumed that the particles are so small they behave just like a gas. Stated in the report is that the particulate size distribution in the engine exhaust is significantly less than 10 micrometers making this test procedure valid due to collection size of PM captured during sampling.

Table 27. Listing of Engines Tested

Engine Test Location	Engine	Flow Determination Method
Kelly AFB	T56-A-7	Calculated by carbon balance, tracer gas and F-factor.[1]
	TF39-GE-1C	Calculated by carbon balance, tracer gas and F-factor[1]
	GTCP85-180	Measured using EPA Methods 1-4
	GTCP-165-1	Measured using EPA Methods 1-4
Corpus Christi Army Depot	T700-GE-700	Measured using EPA Methods 1-4
Laughlin AFB	J69-T-25	Calculated by carbon balance, tracer gas and F-factor[1]
	J85-GE-5A	Calculated by carbon balance, tracer gas and F-factor[1]
Tinker AFB	F110-GE-100	Calculated by carbon balance, tracer gas and F-factor[1]
	F108-CF-100	Calculated by carbon balance, tracer gas and F-factor[1]
	TF33-P-7/7A	Calculated by carbon balance, tracer gas and F-factor[1]
	F101-GE-102	Calculated by carbon balance, tracer gas and F-factor[1]
	TF33-P-102	Calculated by carbon balance, tracer gas and F-factor[1]
Charleston AFB	F117-PW-100	Calculated by carbon balance, tracer gas and F-factor[1]
Edwards AFB	F118-GE-100	Calculated by carbon balance, tracer gas and F-factor[1]
	F404-GE-F102/400	Calculated by carbon balance, tracer gas and F-factor[1]
	F110-GE-129	Calculated by carbon balance, tracer gas and F-factor[1]
	F100-PW-100	Calculated by carbon balance, tracer gas and F-factor[1]
	F100-PW-229	Calculated by carbon balance, tracer gas and F-factor[1]
Naval Aviation Depot, Cherry Point, NC	T64-GE-100	Measured using EPA Methods 1-4
Barnes ANGB	TF34-GE-100A	Calculated by carbon balance, tracer gas and F-factor[1]

estimated with calculations. based on tracer gas methodologies and a carbon balance[12]. Measured ambient air concentrations of PM were subtracted from the measured source concentrations. It should be noted that ambient concentrations were much less (approximately one seventh) of the measured source concentrations.

Testing was accomplished at standard power settings, typically three to five actual flight settings, which included the modes of idle (I), approach (A), intermediate (N), military (M), and afterburner (AB). Table 28 lists the modes tested for each engine. It should be noted that although testing was done for all engines, PM was not measured for the F100-GE-129, F100-PW-100, and F100-PW-229 engines. Because of the way auxiliary power

---

[12] A carbon balance is a mass balance of carbon entering the engines as fuel, and leaving in the exhaust. The exhaust is primarily carbon dioxide and carbon monoxide, with carbon black being much smaller.

units are used, only one power setting was tested. The helicopter engines were tested at four power settings.

Many tests were performed. The data were summarized and averages of the individual tests were presented. Table 29 lists averages for the filterable PM and total PM (total PM includes condensable or liquid PM). The values in Table 29 provide emission indexes in two forms: pounds per hour of operation and pounds per 1000 pounds of fuel burned. A summary of the engine operating conditions including mode, fuel flow, torque (for one engine), shaft horsepower (for one engine), percent of maximum horsepower (for one engine), average thrust, and percent of maximum thrust is shown in Table 30.

4.0 CONCLUSIONS AND RECOMMENDATIONS

4.1 General Conclusions

Important findings on PM related to the exhaust of jet engines include:

- Small PM may be a health concern.
- It is a good approximation that all PM emitted by modern transport aircraft has an aerodynamic diameter of less than 2.5 micrometers. This is an important concern and controlled by the EPA health-based standards for $PM_{2.5}$ as well as PM_{10}.
- The EPA PM standards are massed based (mass/volume of air) at receptor locations but the most complete data base for transport aircraft is the ICAO database which reports the smoke number. The smoke number does not correlate well with mass emissions due to the nature of the test procedure used. As such, there is a lack of measured data to assist in the analysis to determine if an airport is in compliance with the EPA standards.
- PM are irregular in shape and often coagulate. This coagulation process results in different PM characteristics for different age plumes. This leads to a bi-modal distribution. A lognormal distribution is still appropriate for the soot component (non-volatile PM primarily containing carbon).
- PM include both volatile and non-volatile components. Soot is the most prevalent, non-volatile component. Metals are emitted, but in extremely small amounts.
- Effects on PM emission indices include fuel flow, engine design / operating conditions, altitude, and fuel composition.
- Efforts to predict emission indices, or more specific emission factors, may be characterized into four groups: simple factor, compound factor, grab samples or nearby measurements, and measurement based factors.

Table 28. Engine Power Settings Sampled [EQM, 1998]

ENGINE	POWER SETTINGS SAMPLED
T56-A-7	I, A, N, M
TF39-GE-1C	I, A, N, M
GTCP85-180	Single constant setting
GTCP165-1	Single constant setting
T700-GE-700	See paragraph below
F110-GE-100	I, A, N, M, AB (Zone 1)
F101-GE-102	I, A, N, M, AB (Zone 1)
TF33-P-102	I, A, N, M
F108-CF-100	I, A, N, M
TF33-P-7/7A	I, A, N, M
J69-T-25	I, A, N, M
J85-GE-5A	I, A, N, M
F117-PW-100	I, A, N, M
F118-GE-100	I, A, N, M
F404-GE-F1D2/400	I, A, N, M, AB (Zone 3)
F110-GE-129	I, A, N, M, AB (Zone 1)
F100-PW-100	I, A, N, M, AB (Zone 1)
F100-PW-229	I, A, N, M, AB (Zone 1)
T64-GE-100	See paragraph below
TF34-GE-100A	I, A, N, M

Table 29. Particulate Matter Emission Summary [EQM, 1996]

Engine/Setting	Fuel Flow lbs/hr	Filterable Particulate lbs/hr	Filterable Particulate lbs/1,000 lb fuel	Total Particulate lbs/hr	Total Particulate lbs/1,000 lb fuel
T56-A-7					
Idle	724	1.27	1.75	2.63	3.64
Approach	880	1.47	1.67	3.38	3.85
Intermediate	1742	1.57	0.90	2.54	1.46
Military	2262	2.01	0.89	2.76	1.22
TF39-GE-1C					
Idle	1448	0.59	0.41	4.05	2.77
Approach	10477	7.92	0.75	12.52	1.19
Intermediate	12541	6.47	0.52	11.15	0.89
Military	13862	5.77	0.42	16.40	1.18
GTCP85-180 (APU)					
Constant Setting	270	0.15	0.55	0.19	0.72
GTCP165-1 (APU)					
Constant Setting	273	0.09	0.35	0.13	0.48
J69-25					
Idle	167	0.28	1.68	0.53	3.16
Intermediate	872	0.47	0.54	0.82	0.93
Military	1085	0.32	0.29	0.73	0.67
J85-5A					
Idle	434	0.35	0.68	2.40	4.70
Intermediate	950	0.88	1.10	1.43	1.79
Military	2740	2.68	1.08	2.79	1.13
Afterburner (Zone 1)	8138	1.26	0.16	1.93	0.25
F110-GE-100					
Idle	1111	1.65	1.49	2.89	2.61
Approach	5080	2.34	0.46	6.94	1.37
Intermediate	7332	1.22	0.17	4.22	0.57
Military	11358	1.58	0.14	1.58	0.14
Afterburner (Zone 1)	18088	6.72	0.37	60.57	3.34

Table 29. Particulate Matter Emission Summary (continued) [EQM, 1998]

Engine/Setting	Fuel Flow lbs/hr	Filterable Particulate lbs/hr	Filterable Particulate lbs/1,000 lb fuel	Total Particulate lbs/hr	Total Particulate lbs/1,000 lb fuel
F108-CF-100					
Idle	1136	2.17	1.91	2.35	2.07
Approach	2547	2.02	0.79	3.95	1.55
Intermediate	5650	1.64	0.29	3.66	0.65
Military	6458	3.66	0.57	10.27	1.59
TF33A-P-7					
Idle	1093	2.54	2.33	6.69	6.13
Approach	4884	10.81	2.21	17.95	3.68
Intermediate	6356	22.65	3.57	33.59	5.29
Military	8264	19.35	2.34	29.55	3.58
F101-GE-102					
Idle	1117	1.36	1.21	2.43	2.17
Approach	4533	2.15	0.47	19.10	4.23
Intermediate	6557	4.10	0.63	8.84	1.35
Military	7828	3.68	0.47	13.11	1.68
Afterburner (Zone 1)	15314	7.11	0.46	43.87	2.86
T700-GE-700					
Idle	134	0.07	0.51	0.20	1.48
Flight Idle	469	0.56	1.19	0.59	1.26
Flight Max	626	0.81	1.29	1.39	2.22
Overspeed	725	1.01	1.39	1.89	2.60
TF33-P-102					
Idle	1114	1.00	0.90	5.53	4.98
Approach	4737	8.98	1.90	16.82	3.55
Intermediate	5782	9.99	1.73	18.22	3.15
Military	7561	11.28	1.49	19.02	2.52
F117-PW-100					
Idle	978	1.88	1.90	10.43	10.54
Approach	4645	2.00	0.43	25.69	5.52
Intermediate	10408	9.32	0.90	24.06	2.31

Table 29. Particulate Matter Emission Summary (continued) [EQM, 1998]

Engine/Setting	Fuel Flow lbs/hr	Filterable Particulate lbs/hr	Filterable Particulate lbs/1,000 lb fuel	Total Particulate lbs/hr	Total Particulate lbs/1,000 lb fuel
F118-GE-100					
Idle	1097	0.23	0.21	1.37	1.25
Approach	3773	8.99	2.41	17.73	4.47
Intermediate	6350	2.08	0.19	19.37	1.78
Military	10887	1.76	0.16	17.89	1.64
F404-GE-F1D2/400					
Idle	685	0.94	1.37	3.06	4.48
Approach	3111	1.81	0.58	4.53	1.46
Intermediate	6464	4.35	0.67	10.17	1.57
Military	7739	5.58	0.72	12.48	1.61
Afterburner (Zone 3)	15851	5.75	0.36	56.55	3.57
T64-GE-100					
Ground Idle	298	0.06	0.21	0.70	2.36
75% Normal	941	1.43	1.52	1.85	1.96
Normal	1698	1.24	0.73	2.73	1.60
Military	1848	1.53	0.83	1.69	
TF34-GE-100A					
Idle	498	2.26	4.38	4.05	8.00
Approach	933	3.82	4.09	5.79	6.19
Intermediate	1512	2.99	1.98	13.50	8.93
Military	2628	2.58	0.98	6.99	2.67

Table 30. Engine Operation Summary [EQM, 1998]

Engine Type	Operation Mode	Fuel flow, lbs/hr	Torque, Inch-Pounds	Shaft Horsepower	% Maximum Horsepower	Average Thrust, lbs	% Maximum Thrust
T56-A-7	Idle	723.6	1,011	217	4.7	...	...
	Approach	880.2	3,231	688	15.0	...	...
	Intermediate	1,741.9	12,802	2,808	61.2	...	...
	Military	2,261.7	18,754	4,115	89.6	...	...
TF39-GE-1C	Idle	1,448.3	...	...	...	2,955	7
	Approach	10,477.4	...	...	...	31,880	76
	Intermediate	12,541.3	...	...	...	36,617	87
	Military	13,861.8	...	...	...	39,486	94
GTCP85-180	Constant	270.3	100	67	(a)	...	...
GTCP165-1	Constant	272.6	217	132	(a)	...	...
T700-GE-700	Ground Idle	134	384	62	3.8	...	...
	Flight Idle	469	2,700	906	55.9	...	...
	Flight Max	626	4,008	1,333	82.2	...	...
	Overspeed	725	4,848	1,620	99.8	...	...
J69-T25A	Idle	167	...	...	...	73.22	4.3
	Intermediate	872	...	...	...	643	62.7
	Military	1,085	...	...	...	864	84.3
J85-GE-5A	Idle	434	...	...	...	97	3.6
	Intermediate	950	...	...	...	400	15.0
	Military	2,740	...	...	...	2,349	88.0
	Afterburner	8,138	...	...	...	3,310	116.0
F108-CF-100	Idle	1,136	...	...	...	1,990	9.2
	Approach	2,547	...	...	...	6,591	30.5
	Intermediate	5,650	...	...	...	15,123	69.9
	Military	6,458	...	...	...	16,978	78.5
TF33-P-7/7A	Idle	1,093	...	...	...	814	3.9
	Approach	4,884	...	...	...	9,349	44.5
	Intermediate	6,356	...	...	...	12,236	58.3
	Military	8,264	...	...	...	15,349	73.1
F101-GE-102	Idle	1,117	...	...	...	892	5.0
	Approach	4,533	...	...	...	8,143	47.0
	Intermediate	6,557	...	...	...	11,507	66.0
	Military	7,828	...	...	...	13,477	77.0
	Afterburner	15,314	...	...	...	18,460	106.0
TF33-P-102	Idle	1,114	...	...	...	976	5.4
	Approach	4,737	...	...	...	8,783	48.9
	Intermediate	5,782	...	...	...	10,676	59.3
	Military	7,561	...	...	...	13,551	75.3
F110-GE-100	Idle	1,111	...	...	...	592	3
	Approach	5,080	...	...	...	7,645	44
	Intermediate	7,332	...	...	...	11,595	66
	Military	11,358	...	...	...	17,460	100
	Afterburner	18,088	...	...	...	19,780	113
F117-PW-100	Idle	978	...	...	...	1,478	4
	Approach	4,645	...	...	...	13,088	31
	Intermediate	10,408	...	...	...	28,526	68
F118-GE-100	Idle	1,097	...	...	...	NA	NA
	Approach	3,773	...	...	...	NA	NA
	Intermediate	6,350	...	...	...	NA	NA
	Military	10,887	...	...	...	NA	NA

(con't)

Table 30. Engine Operation Summary (continued) [EQM, 1998]

Engine Type	Operation Mode	Fuel flow, lbs/hr	Torque, Inch-Pounds	Shaft Horsepower	% Maximum Horsepower	Average Thrust, lbs	% Maximum Thrust
F404-GE-F1D2/400	Idle	685	---	---	---	632	6
	Approach	3,111	---	---	---	4,057	38
	Intermediate	6,464	---	---	---	8,305	79
	Military	7,739	---	---	---	9,608	91
	Afterburner	15,851	---	---	---	12,034	114
F110-GE-129	Idle	961	---	---	---	809	4
	Approach	4,832	---	---	---	8,034	45
	Intermediate	6,939	---	---	---	11,431	65
	Military	8,611	---	---	---	13,489	76
	Afterburner	15,564	---	---	---	17,467	99
F100-PW-100	Idle	1,067	---	---	---	1,174	8
	Approach	2,726	---	---	---	3,963	27
	Intermediate	7,549	---	---	---	10,992	75
	Military	9,211	---	---	---	12,827	87
	Afterburner	12,198	---	---	---	13,909	95
F100-PW-229	Idle	1,087	---	---	---	806	5
	Approach	3,098	---	---	---	3,768	21
	Intermediate	5,838	---	---	---	8,771	49
	Military	11,490	---	---	---	15,382	86
	Afterburner	20,793	---	---	---	18,218	102
T64-GE-100	Ground Idle	298	1,284	85	2	---	---
	75% Normal	941	6,564	1,458	34	---	---
	Normal	1,698	15,816	3,521	81	---	---
	Military	1,848	17,580	3,873	90	---	---
TF34-GE-100A	Idle	498	---	---	---	665	7
	Approach	933	---	---	---	2,550	28
	Intermediate	1,512	---	---	---	4,200	46
	Military	2,628	---	---	---	7,100	78

(a) Maximum horsepower not available for the auxiliary power units.
--- Blanks indicate a parameter which is not monitored during operation in the test cell.
NA - Thrust values were not available for this engine.

4.2 Suggested First Order Approximation

4.2.1 Background

The literature review has permitted a first order approximation method for PM to be suggested for commercial transport operations. This suggested approximation method is based on several key considerations. The considerations include:

- The Airport modeling community needs to account for changes in fleet mix, aircraft modes (related to throttle settings), and airport altitudes. The simple approximation method and the grab-sample / deposition methods do not permit this flexibility. Accordingly, these methods do not meet current requirements and were not considered further.
- The accuracy of each possible method and the availability of data also were heavily weighed when considering this approximation method. It is a foregone conclusion that measured data would be more accurate than estimation techniques. However, insufficient information exists now or is expected to be developed in the near future to support an entirely new measurement methodology.
- The only comprehensive database now available is the ICAO listing of smoke numbers, which are not well related to mass emissions.
- The compound factor approach has been used by the airport modeling community and could provide the short-term, first-order approximation that is needed. The largest source of uncertainty in this method is correlation between mass emissions and smoke number. To help reduce the potential error due to the uncertainty of the correlation, the compound factor method must use an adaptation of methodologies that have been derived based on the limited amount of existing measured data which correlate mass and smoke number. The suggested methodology is a combination of the methodologies put forward by UMR and DLR. This combined method should allow a more emission index for PM to be derived for use in the compound factor method.

If we again consider the compound factor general format that was previously shown (Equation 2):

$$EI_i = (SN_i / SN_{ref})(EI_{ref}) \qquad\qquad [2]$$

This method must be altered. Instead of a ratio of the smoke numbers and a reference emission index, this index would be based on curve fitting techniques from the limited amount of existing measured data. Consider Figure 6, the work of Champagne. Here a non-linear relationship was shown for the SN to mass concentration. This more closely reflects the SN test method which is based on light reflectance rather than total mass capture. This method was further developed by DLR using the data of Whyte and Hurley to adjust the original curve presented by Champagne. UMR provided further insight by suggesting that only a few categories of aircraft were needed since so many airframes use common engines. UMR put forward the idea that the overall PM index could be related to fuel flow.

If we combine these ideas, and the measured data available, a curve such as that developed by DLR can be derived, but with the additional step of relating to fuel flow as done by UMR. In other words, a specific emission index could be derived that would be both aircraft specific from the individual smoke numbers and related to fuel flow in the ICAO database. Additionally, since these smoke numbers are also presented by mode in many cases, the mode influence could also be considered. As such, the objectives of the first order approximation would meet immediate needs.

4.2.2 Proposed Method

Consider Figure 19. This figure shows the trends as reported by DLR [Petzold, 1998] from the data of Champagne [Champagne, 1971], Whyte [Whyte, 1982] and Hurley [Hurley, 1993].

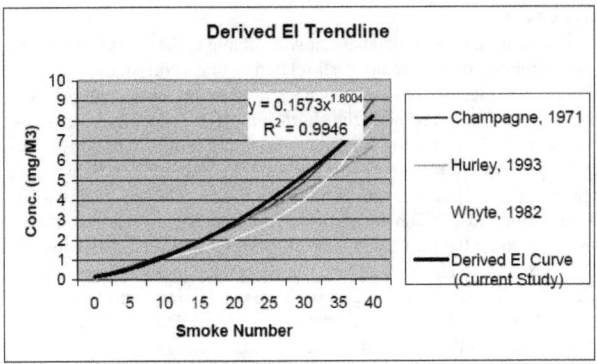

Figure 19. Derived Trendline

If these data are assumed to be representative of commercial aircraft operations, then a derived trend can be determined as also shown in Figure 19. It should be noted that the trendline was purposely derived to provide an upper limit to the presented data and as such is considered conservative. Also, as shown in the figure, a power law equation, with an extremely good fit to the data can be determined. By use of the reported ICAO smoke numbers by mode, modal emissions are considered when smoke numbers are provided for each model. In some cases, the ICAO database only reports the maximum smoke number. With no additional information, the analyst must use this single value for all modes. This will be a very conservative approach since greater estimation of the PM mass will occur. Then, to relate to fuel flow, one more step is needed.

In the next step, we assume the nominal temperature of the flow during smoke number testing is 160 degrees Centigrade. This is a temperature selected from the middle of the required temperature range as outlined by ICAO, Annex 16 [ICAO, 1993]. At this temperature, and at one atmosphere of pressure, one mole of an ideal gas would occupy a volume of 3.062×10^{-2} cubic meters. If it is then assumed that jet fuel can be approximated by the stoichiometric chemical expression put forward by Underwood [Underwood, 1996] then each kilogram of fuel burned would cause 138.554 moles of exhaust to be emitted. This equates to an exhaust volume of 4.925 cubic meters for each kilogram of jet fuel burned at stoichiometric conditions. The product of the volume and the derived concentration equation of Figure 19 yields the desired emission index for predicting the mass of PM for commercial aircraft as follows:

$$EI_{Mass\ of\ PM} = 0.775\ (SN)^{1.8}\ (FF) \hspace{4cm} [8]$$

Where:

$EI_{Mass\ of\ PM}$ = emission index: mg of PM emitted per second
SN = the ICAO reported smoke number
FF = the ICAO reported fuel flow by mode in kilograms/sec

The product of the emission index presented and the time-in-mode would result in a mass based approximation and follow the general method used for other pollutants in the ICAO database. With the derived, aircraft-specific emission factor, the total mass for emission inventories would be derived as before:

$$M_{total} = \Sigma_i\Sigma_j(EI_{Mass\ of\ PM})(N_i)(N_{ei})(t_{mode\ i}) \hspace{3cm} [9]$$

Where:

M_{total} = total mass emitted in mg
N_i = the number of aircraft evaluated
N_{ei} = the number of engines per aircraft type i
$t_{mode\ i}$ = the time-in-mode for each aircraft type i for each mode j

Of course the major limitations to this model are that small particles are not well represented by the smoke number, the combustion process varies by engine design, and the fuel-to-air ratio will change with each mode. Additionally, if only the maximum SN is provided in the ICAO data base, over-estimation could occur. Continual improvements need to be made to this method. Regardless of these limitations, the derived mass-based factor should be more accurate than those that have been used in the past.

4.2.3 Reasonableness Test

A reasonableness test of the proposed method was conducted. The results of one set of tests are presented in Figure 20. In this figure, the derived method is compared to the measured non-volatile PM mass emissions that were reported by Whitefield (Whitefield,

2001) for an RB211-22B engine. It should be noted that these data were not used in the derivation and as such may be considered independent data. The results are quite promising. However, other comparisons have been less promising, showing differences by more than two orders of magnitude. A conservative approach was taken, but in some cases the reported measured mass is greater. Many variables could cause this difference from the predicted to the measured levels, with only speculation of the reasons at this time due to a lack of data. This implies that further refinements and additional data may lead to better agreement. However, preliminary results indicate that the proposed method deserves considerable merit and may be the best available approach in the short term.

In the longer term, as more measured data become available the model could be further refined and possibly divided into additional aircraft categories, based on engine design (for example the combustor spray nozzles used, pressures generated by compressor section, temperature of combustion, etc). The sparsity and inconsistency of existing data does not support refinement by specific engine type at this time.

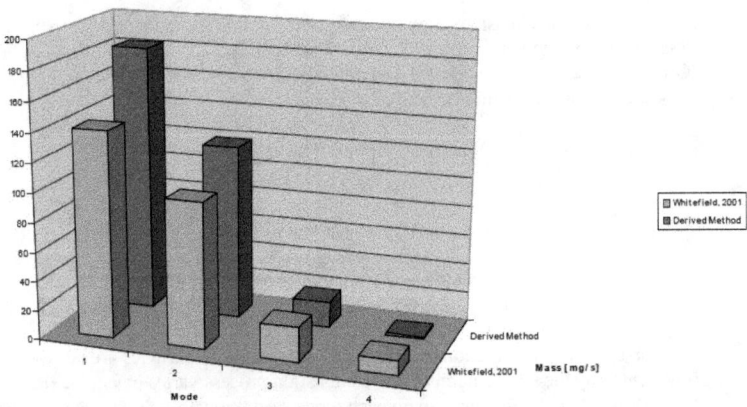

Figure 20. Comparison to Whitefield Data for RB211-22B

Note: Modes shown are:
1. Takeoff
2. Climb-out
3. Approach
4. Idle/Taxi

REFERENCES

[AESO, 1990] Aircraft Environmental Support Office, Summary Tables of Gaseous and Particulate Emissions From Aircraft Engines

[Broderick, 1971] Broderick, A.J., W.E. Harriott, and R.A. Walter, *Survey of Aircraft Emissions and Related Instrumentation*, DOT-TSC-OST-71-5, U.S. DOT, Cambridge, MA., March, 1971.

[Broderick, 1972] Broderick, A.J., M.J. Scotto, and J.C. Sturm, *Particulates in Jet Aircraft Exhaust: Instrumentation and Initial Results*, DOT Transportation Systems Center, Presented at the 65th Annual Meeting of the American Institute of Chemical Engineers, November, 1972.

[Champagne, 1971] Champagne, D.L., *Standard Measurement of Aircraft Gas Turbine Exhaust Smoke*, ASME, 71-GT-88, 1971.

[CFR, 1986], Code of Federal Regulations, *Appendix B – reference method for the determination of suspended particulate matter in the atmosphere (high-volume method)*, CFR 40, Par. 50, 1986.

[Dopelheuer, 1997] Dopelheuer, A., *Berechnung der Produkte unvollstandiger Verbrennung aus Luftfahrttriebwerken*, IB-325-09-97, Inst. Fur Antriebstechnik, Dtsch. Zentrum fur luft- u. Raumfahrt, Koln, Germany, 1997.

[Dopelheuer, 1999] Dopelheuer, A. and M. Lecht, Influence of Engine Performance on Emission Characteristics, *RTO Meeting Proceedings 14 on Gas Turbine Engine Combustion, Emissions and Alternative Fuels*, RTO-MP-14, NATO, Cedex, France, June, 1999.

[Dopelheuer, 2000] Dopelheuer, A. and C. Wahl, *Determination of Quantities and Properties of Aircraft Engine Generated Soot*, Insitut fur Antriebstechnik, Deutshces Zentrum fur Luft- und Ranumfahrt, 2000.

[EPA, 1985] U.S. Environmental Protection Agency, *Compilation of Air Pollutant Emission Factors, Volume 2: Mobile Sources*, EPA/, Office of Research and Development, Washington, D.C., September, 1985.

[EPA, 1995] U.S. Environmental Protection Agency, Determination of particulate emissions from stationary sources, 40CFR60, Appendix A, Washington, D.C., 1995.

[EPA, 1996] U.S. Environmental Protection Agency, *Air Quality Criteria for Particulate Matter (3 Volumes)*, EPA/600/P-95/001, (002), (003), Office of Research and Development, Washington, D.C., April, 1996.

[EPA, 1998] U.S. Environmental Protection Agency, *Profile of the Air Transportation Industry*, EPA/310-R-97-001, Enforcement and Compliance Assurance, Washington, D.C., February, 1998.

[EPA, 1999] U.S. Environmental Protection Agency, *Air Quality Criteria for Particulate Matter (3 Volumes)*, EPA/600/P-99/001, (002), (003), Office of Research and Development, Washington, D.C., October, 1999.

[EQM, 1998] Environmental Quality Management, Inc., *Aircraft Engine and Auxiliary Power Unit Emissions Testing Final Report*, Project Number 3414-088-060, Cincinnati, OH, November, 1998.

[Ernst, 1997] Ernst, D.A. and A.D. Goldman, *Soot Deposition Study: Logan Airport and Surrounding Communities*, KMC Report No. 970114, KM Chng Environmental Inc., Waltham, MA., January, 1997.

[FR,1971] Federal Register, National primary and secondary ambient air quality standards, *FR 36: 8186-8201*, April 30, 1971.

[FR, 1987] Federal Register, Revisions to the national ambient air quality standards for particulate matter, *FR 52: 24634-24669*, July 1, 1987.

[FR, 1997] Federal Register, National ambient air quality standards for particulate matter, final rule, *FR 62: 38652-38752*, July 18, 1997.

[Gardner, 1998] Gardner, R.M., *ANCAT/EC2 Global Aircraft Emission Inventories for 1991/92 and 2015*, report by the ECAC/ANCAT and EC Working Group, Eur. No. 18179, 1998.l

[Hagen, 1992] Hagen, D.E., M.B. Trueblood, and P.D. Whitefield, A Field Sampling of Jet Exhaust Aerosols, *Particulate Science and Technology*, 10:53-63, 1992.

[Hagen, 1996] Hagen, D.E. and P.D. Whitefield, Particulate emissions in the exhaust plume from commercial jet aircraft under cruise conditions, *Journal of Geophysical Research*, Vol. 101, D14, August 27, 1996, pgs. 19,551 – 19,557.

[Hagen, 1998] Hagen, D.E., P. Whitefield, J. Paladino, and M. Trueblood, Particle sizing and emission indexes for a jet engine exhaust sampled at cruise, *Geophysical Research Letters*, Vol. 25, No. 10, May 15, 1998, pgs. 1681-1684.

[Howard, 1996] Howard, R.P, J.C. Wormhoudt, P.D. Whitefield, *Experimental Characterization of Gas Turbine Emissions at Simulated Flight Altitude Conditions*, AEDC-TR-96-3, NASA Lewis Research Center, September, 1996, Arnold Air Force Base, TN., 159 pgs.

[Hurley, 1993] Hurley, C.D., Smoke Measurements Inside a Gas Turbine Combustor, *AIAA 93-2070, 29th Joint Propulsion Conference and Exhibit*, Monterey, 1993.

[ICAO, 1993] ICAO, International Standards and Recommended Practices, Environmental Protection, Annex 16 to the Convention on International Civil Aviation, Volume II, Aircraft Engine Emissions, Second Edition, 1993.

[ICAO, 1995] ICAO, Engine exhaust emission data bank, 1st. Ed., Doc 9646-AN/943, Montreal Quebec, Canada, 1995.

[IPCC, 1999] Intergovernmental Panel on Climate Change, *Aviation and the Global Atmosphere*, Special Report of IPCC Working Groups I and III, Cambridge, UK, 1999.

[Owen, 1992] Owen, M.K., Ensor, D.S., Sparks, L.E., Airborne particles sizes and sources found in indoor air, *Atmos. Environ. Part A*, 26:2149-2162, 1992.

[Petzold, 1998] Petzold, A. and A. Dopelheuer, Reexamination of Black Carbon Mass Emission indexes of a Jet Engine, *Aerosol Science and Technology*, 29:355-356, 1998.

[Petzold, 1999] Petzold, A., A. Dopelheuer, C.A. Brock, and F. Schroder, In situ observations and model calculations of black carbon emission by aircraft at cruise altitude, *Journal of Geophysical Research, Vol 104, No. D18*, September 27, 1999, pgs. 22.171 – 22.181.

[Spicer, 1988] Spicer, C.W., M.W. Holdren, S.E. Miller, D.L. Smith, R.N. Smith, and D.P. Hughes, *Aircraft Emission Characterization*, Report No. ESL-TR-87-63, Battelle Columbus Division, Columbus, OH., March, 1988.

[SAE, 1970] Society of Automotive Engineers, *Aircraft Gas Turbine Engine Smoke Measurements*, Aerospace Recommended Practice, ARP 1179, New York, 1970.

[Underwood, 1996] Underwood, B.Y., C.T. Walke'r, and J. MacKenzie, *Air Quality Implications of Heathrow T5: Calculated PM10 Emissions and Concentrations*, BAA/817, AEA Technology Report No. AEA/16402156/Z/009 Issue 2, London, August, 1996.

[Wayson, 1989] Wayson, R.L. and W. Bowlby, "Airport Air Pollutant Inventories: Pitfalls and Tools," *Transportation Research Record No. 1240*, Transportation Research Board, National Research Council, Washington, D.C., 1989, pgs. 28-36.

[Whitefield, 2001] Whitfield, P.D., D.E. Hagen, G. Siple, J. Pherson, Estimation of Particulate Emission indexes as a Function of Size for the LTO Cycle for Commercial Jet Engines", *Proceedings of the Air & Waste Manangement Association Annual Meeting Orlando*, Florida, June, 2001.

[Whyte, 1982] Whyte, R.B., Alternative Jet Engine Fuels, *AGARD Advisory Report No. 181*, Vol. 2, 1982.

[Yu, 1998] Yu, F. and R. Turco, Contrail formation and impacts on aerosol properties in aircraft plumes: Effects of fuel sulfur content, Geophys. Res. Lett., 25, 1998, pgs. 313-316.

www.ingramcontent.com/pod-product-compliance
Lightning Source LLC
Chambersburg PA
CBHW071758170526
45167CB00003B/1075